3 1257 01467 2066

WITHDRAWN

Schaumburg Township District Library

130 South Roselle Road

Schaumburg, Illinois 60193

THE SOLAR ECONOMY

Renewable Energy for a Sustainable Global Future

Hermann Scheer

WITHDRAWN

EARTHSCAN

Earthscan Publications Ltd
London • Sterling, VA

SCHAUMBURG TOWNSHIP DISTRICT LIBRARY
130 SOUTH ROSELLE ROAD
SCHAUMBURG, ILLINOIS 60193

3 1257 01467 2066

First published in English in the UK and USA in 2002
by Earthscan Publications Ltd

Original title: *Solare Weltwirtschaft*

Copyright © Verlag Antje Kunstmann GmbH, München, 1999

Translated from the German by Andrew Ketley

All rights reserved

ISBN: 1 85383 835 7

Typesetting by PCS Mapping & DTP, Gateshead
Printed and bound in the UK by Creative Print and Design Wales,
Ebbw Vale
Cover design by Andrew Corbett

For a full list of publications please contact:

Earthscan Publications Ltd
120 Pentonville Road, London, N1 9JN, UK
Tel: +44 (0)20 7278 0433
Fax: +44 (0)20 7278 1142
Email: earthinfo@earthscan.co.uk
Web: **www.earthscan.co.uk**

22883 Quicksilver Drive, Sterling, VA 20166-2012, USA

Earthscan is an editorially independent subsidiary of Kogan Page Ltd
and publishes in association with WWF-UK and the International
Institute for Environment and Development

A catalogue record for this book is available from the British Library

Library of Congress Cataloging-in-Publication Data

Scheer, Hermann, 1944-.
    [Solare Weltwirtschaft. English]
    The solar economy : renewable energy for a sustainable global
    future / Hermann Scheer.
       p. cm.
   Includes bibliographical references and index.
   ISBN 1-85383-835-7
    1. Renewable energy sources. 2. Sustainable development. I. Title.

TJ808 .S33 2002
333.79'4—dc21

                                                        2002006934

This book is printed on elemental-chlorine-free paper

Contents

List of figures and tables

Figures

Tables

Preventing climate change: beyond the Kyoto Protocol

'LET'S IMPROVE THE atmosphere' – that was how the German government greeted delegates to the conference on climate change held in Bonn in July 2001, the eighth such conference since 1992. Yet even before the conference took place, it was abundantly clear that even if the Kyoto Protocol were to be implemented in full through to 2012 without being watered down, the most it could achieve would be to bring emissions back down to the already dangerously high levels of 1990. On the basis of existing agreements, the objective was no longer to improve matters, but merely to prevent them getting any worse.

Matters have not been improved by either the discussions in Bonn or the follow-up conference three months later in Marrakech, held to hammer out further details on how the Kyoto Protocol is to be implemented. If implementation were to proceed as planned, the result would be a paltry 2 per cent emissions reduction in those industrialized countries that have signed up. The USA, responsible for 25 per cent of global emissions, would not be taking part. Across the globe, however, total emissions would continue to rise by a further 10 per cent. The gulf between the targets that must be met and the measures that have been agreed is vast. The UN-endorsed Intergovernmental Panel on Climate Change (IPCC) has stated that emissions reductions of 60 per cent by 2050 are vital if the global climate is to be stabilized. There is surely no-one who seriously imagines this can be achieved by prolonging the Kyoto process beyond 2012. The Kyoto debate would appear to have run its course.

In reality, it is now time to open up the debate. When reporting to the public, politicians face understandable pressure to present even minimal results as a success. The truth is, however, that holding international conferences has proved to be an inadequate response to the dangers and challenges that climate change presents. Despite the general consensus that we have to stick to the path originally chosen, it is now past time we asked whether these conferences have not in fact done more harm than good. While the delegates have been debating over the past decade, emissions have been rising by an unprecedented 30 per cent. We can no longer afford to measure the success of climate change conferences in terms of agreements reached. In view of the consensus assumption that such conferences represent the international instrument par excellence for tackling climate change, it is fair to ask how much has been neglected, postponed, cut, omitted or mishandled since they began. The roll-call of failure is so long that it would be irresponsible not to look for a better way forwards. 'Let's improve the policy' should be the new leitmotiv.

At first glance, the case for global climate change conferences appears convincing. Global problems need global – and thus consensual – solutions. All governments must recognize that they have a direct responsibility to tackle climate change, and their commitments must be binding. The right way to achieve such an outcome is to hold global negotiations to decide on a joint programme of action on which no-one can renege. The apparently common-sense nature of this approach, however, is blinding us to basic questions – questions which the now parlous state of the Kyoto Protocol imbues with new urgency. Why should we expect comprehensive, fast and effective policy responses to emerge from what is the most long-winded political decision process imaginable, namely consensus-orientated negotiations between the parties to an international treaty? What were the reasons for the success or failure of other international treaty negotiations? But above all, is it even possible to achieve international agreement on the technological and structural transformation of the energy sector that a successful climate change strategy would require?

The conference process has given governments a perfect excuse to postpone any environmental overhaul of their respective domestic energy sectors until a global treaty has been agreed and ratified, on the pretext that a global framework is essential to preserve international competitiveness. Governments have thus largely been able to forestall taking swifter action at the national level – such as increased taxation on fossil energy – while still protesting innocence on the global stage. The effect of the climate change negotiations has thus been to preserve the status quo. The recent history of the energy industry has seen unprecedented growth in the industry's lobbying power and its ongoing internationalization through forced market liberalization, a process which has received hefty governmental and legislative backing. Movement towards sustainable energy supplies is conspicuous by its absence, and the power of those primarily responsible for global warming is structurally more entrenched than ever. The energy industry's current environmental rhetoric is the only distracting factor in this regard.

National governments have proved themselves incapable of moving on from their traditional role as the protectors of the energy industry at the national level, and they are unlikely to do any better as delegates to international conferences. It comes as no surprise that the most important topics are not even up for discussion: global carbon dioxide taxation; an end to the tax exemption for aviation fuel (although the rapid growth in air travel represents the greatest single danger to the climate); and the abolition of conventional energy subsidies, currently amounting to $300 billion a year. And yet this latter at least would fit nicely with the ideal of free-market capitalism trumpeted by the World Trade Organization (WTO) process.

It is also no coincidence that the global conferences have become fixated on policy instruments such as tradable emissions permits and the win–win solutions that they claim to offer. Environmental economists who front such proposals hope that they can reconcile the interests of the fossil energy industry with the goal of preventing climate change. The energy industry, however, is betting on being able to maintain its established structures and retain its control over global energy investment. These supposedly realistic proposals take on trust the assertion by the energy industry that its interests are identical with those of the economy as a whole, and thus that the costs for individual companies of preventing climate change are burdens on the economy as a whole. Where all the talk is of costs and burdens, it is easy to lose sight of the economic *benefits* of tackling climate change – benefits that will accrue to everybody.

The most important weakness of the Kyoto Protocol, however, is its shaky scientific foundations. The Protocol presupposes that the existing energy infrastructure can be retained; it need only be made more efficient. Tradable emissions credits can be earned only by improving on one link in the chain, namely the ratio of energy input to energy output, for example in a power station or an electric motor. Supply-chain losses before and after the point of conversion are simply ignored. Unrecognized losses and emissions occur in extraction, porcessing, shipping and storage of primary energy, and in waste disposal and distribution. If the efficiency of a power station is increased from 30 to 40 per cent, the final gain over the entire supply chain from extraction to consumption may only amount to an increase from 10 to 12 per cent. Moreover, if a power station in the UK switches from domestically extracted coal to coal imported from Australia, then the lengthened supply chain must necessarily result in increased emissions and energy losses.

Following the liberalization of the global trade in primary energy and the consequent lengthening of global supply chains, it could very well be that the piecemeal calculations set out in the Kyoto Protocol will appear to demonstrate a global emissions reduction, while in fact the opposite has taken place. In other words, the environmental and energy economics of the Protocol has no basis in science. A fossil supply chain can never be truly efficient. The short supply chains of renewable energy sources provide the real key to furture efficiency gains. This is the new paradigm on which this book is founded, and which is invisible to the traditional analyses of the conventional energy system.

Negotiating a global agreement probably only has a real chance of success where the subject of the negotiations is manageable and can be clearly defined, and only a few scattered interests are adversely affected – or when the dominant interest groups expect to benefit on a large scale. The subject of climate change negotiations is the supply and consumption of energy, which is neither manageable nor easy to delineate. And if the benefit in terms of climate protection is to be great enough to justify the considerable international effort, then the interests of the energy industry must inevitably suffer. The outlook for a consensus-based intergovernmental process is consequently less than promising.

By contrast, the Montreal Protocol on the protection of the ozone layer did have a manageable and clearly defined object. The task – difficult enough in itself – was to reign in the interests of certain manufacturers of coolants and cooling systems. The Antarctic Treaty was agreed before any vested interests had arisen, and before any significant investments had been made. The WTO treaty, while extremely broad in scope, matches the interests of the most influential states and other global economic agents. International agreements on disarmament and arms control treaties also have well-defined objects, but go against influential interests in the defence industry. In most cases, unsurprisingly, arms treaties are only ratified if – as in the case of the ban on chemical weapons – the core interests of the defence industry are not significantly affected and the sectors concerned, like the chemicals industry, produce primarily for the civilian market. In other cases, the price of ratification was compensation for the affected interests in the form of new defence contracts in areas not controlled by the respective treaties.

The Kyoto Protocol also contains compensatory measures for the energy industry, and these are not limited to emissions trading and the accreditation of energy-efficient investment in developing countries, but also include the measures agreed in Bonn to compensate the oil-producing countries for lost sales. It is clear, in the light of these so-called 'flexible mechanisms', that the real compromise lies in the widespread failure to consider structural reform of the energy system. The participating countries are tacitly banking on a more efficient fossil energy system, rather than its replacement with renewable energy. Yet the transition to inexhaustible and emission-free sources of energy must form the core of any sustainable climate and environment strategy.

There is no point in constructing a global strategy for climate change if renewable energy is seen as a secondary issue. Where the aim is to replace fossil with renewable energy, there can be no question of compensation for the fossil energy industry. There can be no environmental revolution in energy supply without creative destruction (à la Schumpeter) in the existing conventional energy industry. Renewable energy, correctly understood, must supplant fossil primary energy and the infrastructure and businesses that supply it. Sunlight and wind are supplied by nature free of charge, and biomass primary energy requires a gradual switch from oil, gas and coal suppliers to an entirely different structure of agricultural and forestry businesses. Having set out with the wrong premise, the negotiating parties have been swept along by the ever more absurd logic of the discussions. Their only response has been to build in a system of controls to guard against abuse of the 'flexible mechanisms'. Ever since the decision was taken to pursue climate protection through the instrument of international conferences designed to achieve equitable and binding obligations, it has been inevitable that the goal of climate protection would (at best) be watered down or (more probably) compromised.

It is not just the tangled web of vested interests that makes global climate change negotiations, as they have hitherto been conducted, unlikely to succeed. Even if this web did not exist – and it should be noted that it is broader-based and more intense than the links between politics and the defence industry – there are still economic and technological reasons why a negotiation-based approach has little chance of success. An energy supply that protects the climate and the environment must necessarily be based on renewable, not fossil or nuclear, energy, which means replacing the current system with more efficient energy technology using renewable sources. For this reason, and because renewable energy implies a wholly different supply chain, this is a challenge which calls upon a different set of economic agents to the conventional energy industry and, consequently, it also calls upon other economic interests.

Renewable energy requires a highly distributed approach – each energy consumer is potentially also a producer – while also affording wholly new opportunities for agriculture (biomass), the construction materials industry (energy-efficient materials), engineering professionals and tradespeople (building to make maximum use of the sun), manufacturers of industrial plants, machinery and motors (wind turbines, biogas plants, distributed motor generators, fuel cells), the electrical and electronics industries (devices with no need for mains electricity) and many others besides. Properly followed through, this would be an economic revolution of the most far-reaching kind. The widespread resistance to renewable energy is motivated by fear of the changes this revolution would bring.

History provides many examples of technological revolutions that have reshaped the world. None have run their course without encountering massive resistance; no change has been brought about in consensus with those on the losing end, and none has been the subject of an international treaty, even when its effects were felt on a global scale. Nevertheless, many of these revolutionary changes have needed a political framework or targeted help at their inception in order to develop and showcase the economic and cultural benefits. The list includes railways, electricity grids, the car society, shipping and aviation, nuclear power and telecommunications.

This is the way dynamic processes have developed and continue to develop, to the point where they become self-sustaining (a point which the politically sheltered conventional energy industry has yet to reach). The microelectronic revolution happened because of the productivity gains it brought, despite the almost universal structural upheaval it caused. Countries that promoted microelectronics – for example, through government-sponsored research and development – benefited accordingly. Those who held back in order to forestall economic turmoil subsequently fell behind. The same process can be seen today in the biotech industry.

Demands that these technologies should be introduced on the basis of an international agreement with binding quotas, in order to forestall incalculable economic upheaval, were conspicuous by their absence. Anyone who made such a suggestion would have been derided as an economic illiterate. Countries strove and continue to strive to bolster national competitiveness by being the first to make the next breakthrough. And yet the lessons of the past are comprehensively disregarded in the case of sustainable energy technology, although the range of potential applications is greater than for any other technological innovation.

A dynamic climate change strategy that takes the threat seriously must have at its heart the economic opportunities arising from a revolution in energy supplies. It does not take a global treaty to unlock the benefits of renewable energy. Rather, first one and then ever more states and companies must be prepared to seize new opportunities without pandering to the fossil energy industry. The German Renewable Energy Act leads the way in this respect. To the surprise of international observers, it has resulted in unexpectedly high growth rates and brought forth new industries. Inspired by this example, Egypt, China, India, Brazil, Argentina, France and some US state governments are now developing ambitious wind power programmes of the order of thousands of megawatts.

Trailblazers who proved the doubters and the ignorant wrong were needed to make this happen. Opportunities for such trailblazing are legion, ranging from government research programmes, through agricultural and development policy, to profit-driven entrepreneurial product innovation that has no need of political aid. In the latter case, the greatest opportunities lie in combining microelectronics with photovoltaic technology, what one might call 'solar information technology'. If governments are to put substance behind the climate change rhetoric, then they must fundamentally change their policies on research, agriculture, development aid, architecture and market regulation. Simply plodding on with the intractable Kyoto process and negotiating refinements to the questionable emissions trading policy is not an adequate response.

In future, the primacy of free trade must yield to the more fundamental primacy of active environmental protection if a truly sustainable environmental economy is to be achieved. The global economy can become sustainable only if fossil resources, the consumption of which inevitably gives rise to harmful emissions, are replaced by solar resources that are either emission-free or – as in the case of biomass – whose emissions are naturally recycled by the global ecosystem. Recognizing this truth is a logical consequence of a proper understanding of the laws of thermodynamics. The laws of physics themselves reveal the falsehood of a fossil energy future.

This is not to say that global negotiations have no role to play. Rather, what is needed is a new focus, such as changed priorities for the World Bank, a global renewable energy agency to facilitate technology transfer, reciprocal environmental quality requirements on imports and domestic production, an end to trade restrictions on sustainable energy technology and global standards for the same, a ban on subsidized energy exports and an environmental chamber for the International Court of Justice.

The result would be a dynamic, goal-oriented climate change policy, free of bureaucratic impediments, and a step forwards from simply prolonging and refining the current series of international conferences. Preventing climate change through consensus-building conferences is fantasy politics – all talk and no action.

The Solar Economy offers an alternative programme to the Kyoto Protocol. It details the links between energy resources and economic structures that have given rise to the fossil energy economy, and maps the dynamic road towards renewable energy that will lead to a new and sustainable global economy.

Fossil resources brought the industrialized countries their prosperity. Yet now that their cost outweighs their benefits, fossil resources may bring those self-same countries to their knees. It is the principal thesis of this book that renewable energy, by contrast, brings greater social benefits the more widely it is used, to the point where it fully replaces all fossil energy. There can be no sound reason for making this revolution of our resource base contingent on obligations agreed under international treaties.

Acknowledgements

I WOULD LIKE to thank Heiko Stubner, who worked with my editor Dr Susanne Eversmann on the final version of the text and who compiled the bibliography, and my colleagues Tatjana Brusis, Sigrid Henke, Verena Köln and Karin Schäckeler, for all their help in preparing the manuscript.

List of acronyms and abbreviations

ATM	automatic teller machine
BGR	Federal Institute for Geoscience and Minerals (Germany)
BSE	bovine spongiform encephalopathy
CHP	combined heat and power
CO_2	carbon dioxide
DM	Deutschmark
EEC	European Economic Community
ETSO	Association of European Transmission System Operators
EU	European Union
EUS	Society for Innovative Energy Generation and Storage (Germany)
GATS	General Agreement on Trade in Services
GATT	General Agreement on Tariffs and Trade
GDP	gross domestic product
GENESIS	Global Energy Network Equipped with Solar Cells and International Superconductor Grids
GM	genetically modified/genetic modification
IAEA	International Atomic Energy Agency
ICAO	International Civil Aviation Organization
ICT	internet and communications technology
IEA	International Energy Agency
IIASA	International Institute for Applied Systems Analysis
ILO	International Labour Organization
IMF	International Monetary Fund
IPCC	Intergovernmental Panel on Climate Change
IT	information technology
kWh	kilowatt hours
LDCs	least developed countries

m³	cubic metres
MAI	Multinational Agreement on Investment
MDI	Motor Development International
MW	megawatt
NAFTA	North American Free Trade Area
NATO	North Atlantic Treaty Organization
NGO	non-governmental organization
OECD	Organisation for Economic Co-operation and Development
OPEC	Organization of Petroleum Exporting Countries
PC	personal computer
PV	photovoltaic(s)
R&D	research and development
SADC	South African Development Community
SDP	Social Democratic Party (Germany)
SwFr	Swiss francs
TOE	tonnes of crude oil equivalent
TRIPS	Trade-Related Aspects of Intellectual Property Rights
UNCTAD	United Nations Conference on Trade and Development
UNDP	United Nations Development Programme
V	volts
VCR	video cassette recorder
VDEW	Association of German Electricity Producers
WTO	World Trade Organization

From fossil fuels to solar power: transforming the global economy

Chorus:	*Looking,* *Listening,* *For that which is concealed* *To be revealed,* *Before it is too late* *To put out* *The first few flickers* *Threatening Fire.*
Chorus Leader:	*Many things may start a fire,* *But not every fire that starts* *Is the work of inexorable* *Fate.*
Chorus:	*Other things, called Fate to prevent you* *From asking how they happened,* *Monstrous events,* *Even the total destruction of a city,* *Are mischief.*
Chorus Leader:	*Mischief that wipes out* *Our mortal fellow citizens.*
Chorus:	*Much can be avoided* *By common sense.*

Chorus Leader:	In very truth:
Chorus:	It is unworthy of God, Unworthy of man, To call a stupidity Fate Simply because it has happened. The man who acts so No longer deserves the name, No longer deserves God's earth, In exhaustible, fruitful and kind, Nor the air that he breathes, Nor the sun. Bestow not the name of Fate Upon mankind's mistakes, Even the worst, Beyond our power to put out!
Chorus Leader:	Our watch has begun.

Max Frisch: *The Fire Raisers*, translated by
Michael Bullock

ACCELERATING ECONOMIC AND technological
progress is the hallmark of the modern economic age. Today it
is information technology (IT), biotechnology and genetic
engineering that are developing by leaps and bounds; before
that it was the turn of aeronautics and space travel, atomic
energy in its military and civilian applications, the motor car,
electrification, the railway and the steam-engine. Each of these
new technologies turned existing economic and political structures upside down, and profoundly changed the lives of both
individuals and societies. Their effects are still felt today, with
ramifications that cross physical, geographical, spiritual and
ethical boundaries, the latter most especially where nuclear and
biochemical weapons of mass destruction are concerned. They
also operate on timescales that go beyond our capacity for
responsible action. Nevertheless, as the pace of change continues to accelerate and permeate ever more aspects of our lives,
the modern age is already obsolete. Measured by its claim to
shape the future, it is a thing of the past. The modern age is
already fossilized at heart, built on discards and relics. It has
no real future. We are living in a *fossil economy*.

Today, almost all human activity is critically dependent on energy produced from fossil fuels. Even as the economy scales new technological heights, the energy that powers it is condemning it to death. This fundamental contradiction is no mere Cassandran prophecy, but a truth arising from the operation of the laws of nature. The simple fact is that all economic activity relies on the physical and chemical conversion of materials from one form into another, and the conversion of fuels into the energy needed to distribute and consume the resultant products. Energy and raw materials are the fundament of our economies, their *nervus rerum*, or 'nerve of all things'. This *nervus rerum* is the real 'ghost in the machine' (Arthur Koestler).[1]

The resource base is far more fundamental to economic development than questions of political and social order. The old dispute of capitalism versus socialism pales into insignificance before the life-or-death choice of renewable versus non-renewable resources. It is a peculiarity of the 20th century that debate of this issue has dwindled as the scale – and potential consequences – of energy and resource consumption has escalated. At the beginning of the 20th century, Frederick Soddy wrote in his seminal work *Matter and Energy*:

> '*The laws expressing the relations between energy and matter are not solely of importance in pure science. They necessarily come first... in the whole record of human experience, and they control, in the last resort, the rise or fall of political systems, the freedom or bondage of nations, the movements of commerce and industry, the origin of wealth and poverty and the general physical welfare of the race. If this has been imperfectly recognized in the past, there is no excuse, now that these physical laws have become incorporated into everyday habits of thought, for neglecting to consider them first in questions relating to the future.*'[2]

It is the firm belief that there is no alternative to the fossil-fuel economy that is responsible for reducing the all-important question of energy and resources to the status of a secondary issue. Even economists discuss the energy question only in terms of factors that affect the price level. The availability of energy and resources is taken as given, regardless of source. Where one raw material or energy source is used in place of another, this is regarded as an isolated operational decision that has no intrinsic relevance to the structure of the economy as a whole. Only if additional or reduced costs are involved are there thought to be wider implications.[3] The choice of energy- and resource-base has thus appeared to be a problem for technicians and businessmen – and more recently, for ecologists. This is in keeping with the ideology of the technological era as Jürgen Habermas has described it: the larger context is reduced to its component processes, which can be managed only by specialized, instrumentally rational professionals, and which are no longer a valid subject for wider public debate.[4] There is a tendency to view technological issues in particular as value-free, independent of ideas, interests or the conflicts that arise from their inherent contradictions.

In modern times, the realization that economic activity can have social and political consequences has given rise to the concept of 'political economy'. The 'political economists', however, rarely, if ever, include the natural constraints of physics and technology in their analyses, although, as Hans Immler remarks, 'the industrial wealth, technological progress and the changing shape of our civilization that characterize modern times rely on the productivity of physical and biological ecosystems'.[5] Those in positions of political and economic responsibility lack the knowledge they need of these issues, and the scientists and technicians themselves have lost sight of the wood among the trees of their own specialisms. Now, however, that people have begun to realize that our growing dependence on finite resources may have dangerous consequences for the planet as a whole, and that this dependence has indeed already led to social catastrophe, and with increasing public awareness of the growing dominance of technology, there is a crying need for a new concept of 'political natural economy'.

Our current way of life cannot continue if we remain economically dependent on fossil fuels. It is therefore imperative that we make comprehensive use of solar energy – not just to augment fossil fuels (and with them nuclear power), but to replace them. The global economy owes its better times to the exploitation of fossil fuels – but they will also bring it worse times to come.

The power of the pyromaniacs

The faster the pace of the current global economy, the faster it rushes towards disaster:

- The energy we use in producing, delivering and consuming materials is overwhelmingly derived from fossil sources: crude oil, natural gas and coal, together with nuclear power from uranium.
- Fossil resources – primarily crude oil – and minerals are the most important raw materials for the industrial manufacture of finished and semi-finished goods.

Today's global economy, while proclaiming the ideals of 'open markets' and an 'open society', is thus ultimately a 'closed shop' from which other resources are excluded. The planet on which we live, however, is both an open and a closed system at the same time. It is open to the continual influx of energy from the sun, to the gravitational pull of the sun and the moon and to cosmic rays. It is closed as far as stocks of fossil resources are concerned (at least over the feasible timescales of human activity, the solar origin of these energy sources lying hundreds of millions of years in the past), and with respect to the total quantities of matter, water, land and air available. For as long as the global economy continues to operate on the basis of these limited energy and material supplies, its future prospects will be bleak. There are two incontrovertible reasons for this. Firstly, that supplies of fossil and mineral resources are limited; and secondly, that the processes in which these resources are used inevitably also overstretch, damage and even destroy those limited planetary resources on which our lives depend: the water, the land and the atmosphere.

With respect to energy consumption, this second reason has long since become literally a burning issue. Statistics on world energy consumption show that 32 per cent is generated by burning crude oil, 25 per cent by burning coal, and 17 per cent by burning natural gas. Five per cent comes from nuclear fuels, and another 14 per cent from combustion of biomass – of which only a small proportion is replaced by new planting. Hydroelectricity accounts for a mere 6 per cent of all energy consumed. The use of biomass, which, when combined with parallel new planting, has the potential to become a perpetual source of energy, is current largely confined to the rural hinterland of the so-called developing countries. The global economy as such is 'fired' primarily with crude oil, coal, natural gas and nuclear fuel, and is consequently dependent on the suppliers of these resources. The global economy, and with it the world, is therefore dominated by pyromaniacs intent on burning ever-greater quantities of fossil fuels for as long as they can possibly do so. Despite all the scientists' warnings and the politicians' environmental promises, current trends indicate that world-wide burning of fossil fuels is likely to 'flare up' by 50 per cent between 1990 and 2010 alone.

The world is, as Max Frisch puts it in *The Fire Raisers*, 'favourably situated' for this global pyromania. In his 'Morality Play without a Moral', in response to the question 'What was it you gave them? Did I see right? Were they matches?' Herr Biedermann replies: 'Why not... If they were really fire raisers, do you think they wouldn't have matches?' Like Herr Biedermann, and with the sycophantic approval of politicians, scientists and journalists, the fossil energy industry that supplies its 'fire-power' to all the corners of the Earth disavows all responsibility, pointing instead to the needs of its customers – as if there were no way on this Earth to produce energy without burning nuclear or fossil fuel. This continued – not to mention increasing – dependence on fossil fuels is sending the world's future prospects up in smoke. The global body politic is faced with its most important decision yet. In the final analysis, it is a choice between sunlight and ash.

Fossil fuels will probably run out sooner than mineral resources. Crude oil, natural gas and coal reserves, once burnt, are irrecoverable. Only nuclear waste can be reprocessed to extend its working life as an energy source, although at the cost of increasing the risk of nuclear accident and placing an unacceptable radioactive burden on future generations. Mineral resources, by contrast, are in principle recoverable; finite reserves can thus be extended. However, refining and manufacturing processes are inevitably associated with losses and environmental problems, albeit with varying degrees of severity from material to material.

In general, the danger of ecological destruction resulting from energy generation and manufacturing processes is more immediate than that of the irrevocable exhaustion of resources. It is this consideration which leads Friedrich Schmidt-Bleek, in his endorsement of 'factor 10' – harnessing increased resource productivity to achieve a tenfold reduction in the energy and materials inputs required for the production of goods and services – to argue that it is 'the quantity, not the nature, of the resources employed which is the problem'.[6] The decisive factor according to this argument is not whether the resources consumed are renewable or non-renewable, but whether the manufacturing processes employed are ecologically sound. Similar demands, in the more moderate form of 'factor 4', are also made by Amory B Lovins, L Hunter Lovins and Ernst Ulrich von Weizsäcker.[7] Yet as incontrovertible as the need for increased resource productivity may be, the argument that the choice of energy and material source represents less of a problem than the quantities consumed is one that I utterly reject.

My first proposition is: *global civilization can only escape the life-threatening fossil fuel resource trap if every effort is made to bring about an immediate transition to renewable and environmentally sustainable resources and thereby end the dependence on fossil fuels.* In making this statement, I am not playing off renewable resources against the goal of optimal resource productivity. Any such conflict exists more in the minds of those who play down the potential of renewable resources than in reality. Only with the transition to renewable resources, and thus to a solar global economy, can economic logic and with it the future path of economic development be radically altered. It is this transition which is the key to the future viability of the global economy.

In a global economy based on solar energy, the entire demand for energy and materials can be met from *solar energy sources and solar resources*. The inexhaustible potential of solar, that is to say renewable, energy includes sunlight and solar heating, wind and wave power, hydroelectricity and energy derived from plants and other organic substances. The term 'solar resources' refers to materials of plant origin, produced from sunlight via photosynthesis. Such materials are usually termed 'biomass', renewable or plant-derived raw materials. However, I suggest that the term 'solar resources' should be applied to cover all these materials. Not only does this clearly identify their common origin, it also points the way that can and must be travelled from fossil and other finite energy sources to those which can be produced time and again from the environmentally sustainable source of the sun. By making systematic use of solar resources, ensuring that materials consumed are always replaced by ecologically sustainable new growth, it is possible to meet most of humanity's needs in a way which is sustainable in the long term.

Besides the fundamental environmental reasons, there are also considerations of economics, global security and other social factors which speak for the introduction of solar resources on a large scale:

- In view of the growth imperative in the global economy, the most that increases in productivity and efficiency can achieve is to stabilize resource consumption at its current level, a level which is already higher than either society or environment can sustain. It is therefore essential that productivity goals should be coupled in all cases with a shift to solar resources. This coupling would also result in greater allocative efficiency (that is, a more optimal combination and distribution of investment, productive capacity and materials) than is possible with fossil resources. It is well known that – with the exception of biomass – power generation from solar energy sources is emission-free. What is less well known is that solar energy sources, combined with appropriate power generation technology, also allow resources to be employed in a more tightly focused and productive fashion. In other words, production using solar resources is both less damaging to land, air and water and more efficient in its use of energy. With renewable resources it therefore becomes possible to meet both environmental and economic productivity targets with less effort and, on the whole, in a more cost-effective way.

- As reserves of crude oil, natural gas and certain strategically important minerals approach exhaustion, resource crises are becoming more intense.[8] It is not simply a question of how long and with what environmental consequences we can continue to consume these resources. The location of the reserves is also important: who has economic control, who can set the prices and who, in the end, is able to pay them.

  Disputes over resource access rights can provoke dramatic conflicts. They contain the seeds of true world wars. By comparison, of the two World Wars of the 20th century, the first was confined (in terms of theatres of battle) to Eurasia, and the second largely to Eurasia, North Africa and the Pacific. The 1990–1991 Gulf War and the 1994–1996 civil war in Chechnya are the harbingers of the intensifying struggle for resources, as described by Hans Kronberger in *Blut für Öl* (Blood for Oil).[9] As the curve of falling supply of fossil fuels and strategically important resources draws ever closer to the curve of rising demand from the growing populations of the developing economies, the struggle for control over diminishing conventional resources both within and between continental economic blocs seems set to escalate well before the reserves are finally exhausted. When these two curves intersect, the result will be conflicts more dangerous than any before in world history. But even before this point, crises of availability, price and distribution will have intensified, with unknown consequences for the global economy.

- Energy and mineral resources are found in relatively few locations around the globe, but consumed everywhere. As over time first the industrialized countries and subsequently the world have become dependent on them, energy and mineral resources have been a decisive force in the shaping of political and economic structures the world over. Dependency on these resources had been forcing the 'globalization' of economic activity long before the concept hit the newspaper columns. The drive to seek control of resources has not only steered the policies of the post-colonial nation-states, and latterly the dissolution of the former Soviet Union, but has also determined the foci of economic activity and their attendant economic structures. Whether openly or covertly, resource dependency places societies at the mercy of external factors, thus increasing their susceptibility to crisis.

This book examines the factors which:

- on the one hand, have shaped a global economy which, since the pioneering days of the industrial revolution, has become ever more dependent on supplies of fossil fuels and thereby, despite all our increased technological capabilities, also ever more fragile and in danger of collapse; and
- on the other hand, mean that – and this is my second proposition – *making the groundbreaking transition to an economy based on solar energy and solar resources will do more to safeguard our common future than any other economic development since the industrial revolution.*

The road to the solar global economy will be a rollercoaster ride that touches upon almost all existing interests. There will also be numerous conflicts along the way. In their desire to avoid conflict, many people fear to address the fundamental question of our energy base, or do so only sotto voce, postponing serious discussion until some future date. Yet the longer the global economy remains dependent on fossil energy and mineral sources, the more severe will be the ultimate consequences.

Fossil resource dependency: how economic processes have come adrift from their environmental and social bases

One currently fashionable interpretation has it that mineral and energy resources are playing an ever decreasing role as new and breathtaking technological developments take us forwards into the 'weightless economy' and the 'post-industrial age'. Yet the real legacy of the new technology has been only folly and wilful neglect of the resource issue, reinforcing the illusion that technological solutions can be found for every problem. Already, intelligent machines are gestating in research and development (R&D) departments. According to the euphoric forecasts of American futurologist Michio Kaku,[10] miniaturization, above all in internet and communications technologies (ICT), promises limitless new possibilities and freedoms. In *Metaman*,[11] the biophysicist Gregory Stock enthusiastically describes how the technology of genetic modification (GM) will allow us to incorporate technology into biological processes and thereby partially or totally replace them, and how human and technology will be melded into a monstrous 'superorganism'. One promise of the 'biotech age', as Jeremy Rifkin critically observes,[12] is that our ability to produce foods will become completely free of all natural restrictions. And if we can thus abruptly abandon evolution for a new developmental path, if we can indeed rise to become 'nature's choreographers' (Michio Kaku), then does the world even need the gift of solar resources?

After all, do we not still have enormous reserves of fossil fuels at our disposal, such as the methane bubbles in the ocean bed, or the minerals present in seawater, if we can but learn how to extract them? Will the development of controlled nuclear fusion not solve all our energy problems? And will it not be possible to tap the boundless resources of other planets, or even to open up whole new biospheres? Is the question of resources, which has been raised time and again since the 19th century, and which, according to Wilhelm Fucks, in combination with science and technology constitutes the 'Formula for Power',[13] now therefore redundant? Has the danger of global environmental catastrophe not been shown to be the delusion of jumped-up, technologically illiterate doom-mongers, because the permanent global revolution of technology has rendered all such problems soluble?

Dreams and fantasies the lot. The question of resources is far from obsolete. Anybody who ignores this and places their faith in the technological 'brave new world' (a phrase which, for Aldous Huxley, was a bitter irony, but which the modern techno-pundits now imbue with bright promise) has been dazzled by partial, faddishly exaggerated and overgeneralized reports of the actual developments.

Even if materials are consumed and processed in fully-automated web-enabled factories, the consequences remain the same. The existence of both a 'weightless' or 'post-industrial' economy and increased demand for energy and resources is not a contradiction. The fall in demand for various manufactured products distracts from the fact that aggregate demand can still be increased by the proliferation of energy-intensive service industries, for example, by the rapid growth in transport and tourism. Moreover, demand rises with the growth in world population, and Asia's repetition of the Western fossil fuel industrial model is only just coming into full swing. China and India alone, with two billion inhabitants between them, are home to one third of all humanity. Furthermore, plans to extract materials from ever deeper recesses of the Earth, from the oceans or even from other planets, without regard for the energy costs or the increasing risk to the environment, are in complete denial of reality.

The naive conclusion that the issue of resources does not present a problem (any longer) is in any case refuted by the current redefinition of North Atlantic Treaty Organization (NATO) strategy, which now looks towards the safeguarding of energy and material resources. Military experts are here explicitly admitting the truth that finance ministers and corporations deny.

Global competition in place of global environmental policy

Global energy and manufacturing industries have continued to wreak havoc undisturbed by international agreements to slow down and control development – ie, in spite of the globalization of environmental policy that was initiated with the 'Agenda 21' agreement adopted at the 1992 Earth Summit in Rio de Janeiro. Following publication of the 'Global 2000' report compiled for President Carter at the beginning of the 1980s, the word 'globalization' at first stood for environmental protection.[14] Since then, however, the concept has become a synonym for global competition between businesses, as far as possible unimpeded by import duties or taxation, high wages or socially or environmentally motivated regulations. The legal foundation of this globalization is the treaty establishing the WTO, drawn up in Marrakech in 1994, the purpose of which is to guarantee the largely unimpeded flow of capital, goods and services. The governments which negotiated and signed this treaty had all signed up to Agenda 21 two years previously, yet the contradictions between these two treaties were never a topic for discussion. The WTO rules, by comparison with the vague resolutions on global environmental protection, are fairly specific, binding and even include a system of sanctions for non-compliance. The WTO treaty facilitates and reduces the cost of the transfer and consumption of resources. Its explicitly stated objective of increasing and accelerating trade boosts energy usage in the transport industry; the intended expansion of global trade in agricultural produce promotes the use of environmentally destructive agricultural production methods and widens the scope of activity of the agribusiness firms responsible. The WTO treaty is supposed to enhance economic productivity, but, as a result of the continuing dependency on limited resources and the greater freedom accorded to the highly concentrated extraction industry, its effect is to accelerate the process of destruction.

As things stand today, environmental protection and economic competition are two aspects of globalization that stand diametrically opposed to one another. The freedom of global competition has been declared sacred. It has been accorded a higher political priority than climate protection or conservation of biodiversity. The WTO takes precedence over Agenda 21, competition law over environmental law, the interests of the present over the interests of the future. This divide can only be bridged with a solar resource base. It is not, as many commentators on global environment issues would have it, the wholesale introduction of technology which has led the world into this cul-de-sac, but rather the prevailing resource base and the orientation of technological development and its infrastructure towards fossil fuels. It will not be environmentally motivated Luddism that leads us back out of this blind alley. It will be a firm decision to reject non-solar resources.

My third proposition is that *economic globalization can only be made environmentally sustainable through targeted replacement of fossil fuels by solar energy sources. This is the only way to rein in the destructive imperative of the fossil economy and call a halt to the creeping homogenization of economic structures and cultures. It is the only way to make economic development diverse, sustainable and of lasting benefit to both individuals and society.*

The origins of the fossil-fuel economy

The industrial career of the fossil fuels was launched in the industrial revolution with the invention of the steam-engine, which quickly began to replace human and animal muscle in the productive industries. The 'obsolete' technology of the steam-engine has not been consigned to history: modern nuclear, coal-, gas- and oil-fired power-stations all still work on the same principle and, even today, it continues to shape the structure of the global economy. All the new technologies that have since been developed remain wedded to the same fossil fuel energy base pioneered by the steam-engine.

In its time, James Watt's 1769 invention brought vastly increased energy efficiency, paving the way for the industrial revolution.[15] Greater energy efficiency made mass production possible, with the result that consumption of energy and raw materials grew at a furious rate. Initially, the primary fuel was wood or charcoal. But as the steam-engine became more widely used, demand quickly outstripped the available reserves of wood from nearby forests, and coal became the fuel of choice. The steam-engine's efficiency at converting chemical energy into mechanical work determined the industrial resource base and, once determined, the resource base determined the path of future technological development. Subsequently expanded to encompass crude oil and natural gas, the fossil energy system has been the focus for all subsequent innovation in the field of power generation.

Power generation technologies based on fossil fuels demand high energy densities – ie, high energy content per unit volume, with low transport costs. The easier and the more cost-effectively power could be generated, and the greater the efficiency of the power stations, the more demand grew for what were globally the cheapest sources of energy and materials, and the more the market could be expanded. The industrial revolution became the ever accelerating permanent revolution of the global economy. The fossil energy industry to which the steam-engine gave birth has since become more than a simple driving force: it has made globalization the governing principle for all economic activity.

Accelerating change and global displacement

The speed and scope of globalization had already been documented in Karl Marx and Friedrich Engels's *Communist Manifesto* of 1848. The relevant passage is now more topical than ever. You have only to replace the word 'bourgeoisie' with the modern term 'big business' – albeit they have slightly different characteristics – to arrive at an impressively apt description of the present situation, even down to the one-sided and arrogant conception of economic development:

> *The bourgeoisie cannot exist without constantly revolutionising the instruments of production, and thereby the relations of production, and with them the whole relations of society. Conservation of the old modes of production in unaltered form, was, on the contrary, the first condition of existence for all earlier industrial classes. Constant revolutionising of production, uninterrupted disturbance of all social conditions, everlasting uncertainty and agitation distinguish the bourgeois epoch from all earlier ones. All fixed, fast-frozen relations, with their train of ancient and venerable prejudices and opinions, are swept away, all new-formed ones become antiquated before they can ossify. All that is solid melts into air, all that is holy is profaned, and man is at last compelled to face with sober senses, his real conditions of life, and his relations with his kind.*

The need of a constantly expanding market for its products chases the bourgeoisie over the whole surface of the globe. It must nestle everywhere, settle everywhere, establish connexions everywhere.

The bourgeoisie has through its exploitation of the world-market given a cosmopolitan character to production and consumption in every country. To the great chagrin of reactionists, it has drawn from under the feet of industry the national ground on which it stood. All old-established national industries have been destroyed or are daily being destroyed. They are dislodged by new industries, whose introduction becomes a life and death question for all civilised nations, by industries that no longer work up indigenous raw material, but raw material drawn from the remotest zones; industries whose products are consumed, not only at home, but in every quarter of the globe. In place of the old wants, satisfied by the productions of the country, we find new wants, requiring for their satisfaction the products of distant lands and climes. In place of the old local and national seclusion and self-sufficiency, we have intercourse in every direction, universal inter-dependence of nations. And as in material, so also in intellectual production. The intellectual creations of individual nations become common property. National one-sidedness and narrow-mindedness become more and more impossible, and from the numerous national and local literatures, there arises a world literature.

The bourgeoisie, by the rapid improvement of all instruments of production, by the immensely facilitated means of communication, draws all, even the most barbarian, nations into civilisation. The cheap prices of its commodities are the heavy artillery with which it batters down all Chinese walls, with which it forces the barbarians' intensely obstinate hatred of foreigners to capitulate. It compels all nations, on pain of extinction, to adopt the bourgeois mode of production; it compels them to introduce what it calls civilisation into their midst, ie, to become bourgeois themselves. In one word, it creates a world after its own image.

The bourgeoisie has subjected the country to the rule of the towns. It has created enormous cities, has greatly increased the urban population as compared with the rural, and has thus rescued a considerable part of the population from the idiocy of rural life. Just as it has made the country dependent on the towns, so it has made barbarian and semi-barbarian countries dependent on the civilised ones, nations of peasants on nations of bourgeois, the East on the West.

The bourgeoisie keeps more and more, doing away with the scattered state of the population, of the means of production, and of property. It has agglomerated production, and has concentrated property in a few hands. The necessary consequence of this was political centralisation. Independent, or but loosely connected provinces, with separate interests, laws, governments and systems of taxation, became lumped together into one nation, with one government, one code of laws, one national class-interest, one frontier and one customs-tariff.[16]

The search for unlimited profitability growth and the economic laws of capital employment, which lead to productivity growth, industrial concentration, market expansion and the struggle to eclipse competitors, did not originate in the Industrial Revolution. They are not the exclusive preserve of the bourgeoisie, nor of modern managers; they will probably always be with us. But the choices we face are not always the same.

It is of course by no means just the 'bourgeoisie' who have trodden this path. All those whose livelihoods depend, or who feel dependent, on the world order created by the Industrial Revolution and subsequent technological revolutions, even left-wing political parties and trade unions, have long since counted its defence among their vital interests. As Jan Ross, writing in *Die Zeit*, observes,[17] the highly organized institution that is international capitalism can manage without the bourgeoisie and without unincorporated risk-taking entrepreneurs. In fact, the functionaries in the corporate upper echelons even regard longer-term responsibilities as an unwelcome intrusion. Too many cultural or social scruples are an impediment to the smooth running of global business. States which in the 20th century were without a bourgeoisie, such as the USSR, have nevertheless followed the same techno-economic imperatives of the Industrial Revolution through central planning, although the giant energy and mineral reserves of the USSR in particular removed the need to establish global supply chains. The failure of the socialist economic experiment was probably due to the fact that the USSR attempted to follow the same path of industrial development as its Western politico-economic adversary, but with bureaucratic inefficiency in place of entrepreneurial drive. Following its collapse, the former superpower has been folded into the global economy, its energy and mineral reserves opened up to the global market.

Accelerated by modern technology, the economic displacement of the past 200 years is now reaching full speed. The first phase was the displacement of the so-called primary sector of agriculture and forestry, which before the Industrial Revolution had employed more than three quarters of the population, by the so-called secondary sector of industrial manufacturing. In 1900, manufacturing employed the majority of the workforce. Then came the explosion in the so-called tertiary or service sector, which mopped up those made redundant by manufacturing productivity growth. In the year 2000, the majority of the workforce in the industrialized countries was employed in the tertiary sector. This process has since been repeated at various intervals the world over, the socialist centrally planned economies included. The displacement of salaried work, however, has by no means come to an end. Now it is information technology which is penetrating all sectors, facilitating the replacement of human labour by fossil fuel energy and technology in the rumps of the primary, secondary and tertiary sectors.

The faster the process of displacement, the greater the social upheaval caused. Where attempts have been made to skip sequential phases of sectoral development, the consequences have normally been disastrous – above all in the developing countries. Industrial concentration has been the driving force behind this process, culminating in the 'corporate empires' of the transnational conglomerates. These conglomerates are now actively seeking to remould state institutions to suit their interests in an ever more direct and blatant way. One state will be played off against another, and democratically elected governments are becoming the puppet regimes of the corporate giants. It is no accident, as this book will show, that most of these 'global players' belong to the energy and resource extraction industries: the energy, mining and agribusiness corporations.

International institutions must be strengthened to mitigate the influence of global corporations. Corporations, however, are better endowed, more influential, more sure of their goals, more flexible, more effective and better organized than any institution could be. They merge according to need; they know how to harness the political and scientific elite to achieve a global consensus. They can force acceptance of an international economic order favourable to business, but which deprives democratic governments of vital rights to shape their own economic policy, thereby weakening their ability to discharge their essential responsibilities.[18] They acknowledge no responsibility for the future, nor for human or environmental welfare outside their own business areas – that is, unless motivated by a sense of moral obligation, as part of an advertising and marketing strategy or through charitable activity or one-off donations. Transnational corporations are well on the way to erecting a privately run global planned economy in the form of global cartels. In heedlessly following the logic of their own specific constraints they are bringing about a back-to-front version of the Marxist utopia: capital and corporations are internationalizing, but not – or if so, to a lesser extent – their dependants. The state is abolished, not in favour of free communities, but in favour of private business organizations. The hallmark of the economically globalized world is not 'liberté, égalité, fraternité', which since the French Revolution have been the stated ideals of humanity and democracy, but rather the creeping redundancy of democratic institutions and the widening gulf between rich and poor. The environmental slogan 'think globally – act locally' has been taken to heart by the corporate empires: act globally and profit locally. Whoever can command global resources effectively rules over nature and, in the end, over nations and their governments.

Business unbound: cutting loose from nature and society

Right from the dawn of the Industrial Revolution, businesses have been systematically cutting loose from their geographical, social, cultural and environmental bases – and, in the world of currency and financial speculation, from even their entrepreneurial basis. The world of fuel and mineral extraction has come adrift from the world of power generation and manufacturing; manufacturers have lost touch with their markets, and seed production has been divorced from agriculture. Pollution and polluters are increasingly removed from the places where their destructive effects are felt. Democratically controlled political institutions are also being gradually cut off from what are increasingly international decision-making forums. The decisions of today have less and less to do with prospects for the future. People are being cut off from their culture; the humanitarian values they have been taught to respect are divorced from the realities of daily life. The machinery of global business is accelerating these developments, leaving no place for rest or security and urging people to be ever more ruthless, even, ultimately, to themselves. This global machine is operating far beyond the margins of safety.

Criticism of this kind of globalization has been intensifying but, at the same time, there is growing helplessness in the face of the question of how it can be directed along socially and environmentally sustainable lines. Although new forms of sustainable business are beginning to arise, the rate of take-up lags behind the pace of destruction. Social compensation for the upheavals in the global economy can no longer match the speed at which they occur. Political institutions struggle to keep up, while at the same time their scope for action shrinks – until, exhausted, they drop out of the race, either redefining their responsibilities or relinquishing all claim to political authority.

All sides agree that local business structures are indispensable. Retaining their viability against global market forces and their global corporate flagships, however, has become a difficult and expensive exercise, one which now seems hopeless. The same goes for the support, generally recognized as necessary, for more labour-intensive small and medium-sized enterprises faced with waves of global mergers. Of course, it is still important to seek to influence the business cycle, to use demand-side policies to stabilize domestic markets, to reform social security systems, to encourage greater individual responsibility, to reduce working hours in order to distribute the work more widely, to provide incentives for investment and open up new opportunities for growth, to promote the foundation of new businesses and dismantle obsolete bureaucratic restrictions, to construct unified economic institutions across continents, to establish a global political framework to bring speculative gambling in the international stock and money markets under control – and to put together the right package of policies to suit all these aims.[19] Yet with these measures, political decision-makers can only mitigate the negative social and environmental effects of mainstream developments; overcoming the globalization process which encompasses and permeates all sectors of the economy is beyond their reach. Environmental dangers in particular cannot be averted by corrective measures that only operate within the existing framework of the fossil-fuel economy.

But can we not rely on the development, manufacture and marketing of new products to create new jobs? For decades, whenever an unmet need could be found or stimulated in the mass market, this strategy has produced boom after boom, until the relevant market was saturated – the famous Kondratieff cycles. We can expect new mass-market products to appear, just as we can expect further boom cycles. However, on the high-speed lines of the international business world, the booms to come will not bring new sources of mass employment. The experience of the IT revolution has been just that: new industries have been created, but the overall result for the economy has been not an increase in job opportunities, but a reduction. A similar situation can be seen in the biotech industries: new companies and new jobs in the relevant industries, but against these new jobs must be set massive job losses in agriculture worldwide. Those who argue otherwise are taking an arbitrarily narrow view of events in the global economy. It is the cycles that matter, in economics as in nature.

Reconnecting business and society
through solar resources

It is not just politicians, businesspeople and trade unionists who are at a loss in the face of the rising tide of globalization, but also the economists. As it has become clear that the rise of the modern global economy, especially in combination with today's global business structures, has set innumerable time bombs for future generations, the necessity of a sea change in economic behaviour has become obvious. The neoliberal approach claims to bring just such a change. Yet there is growing recognition that it is doomed to failure. The neoliberals cannot deliver on their promises because, among other reasons, they either cannot or will not act against market domination and market distortion, and they draw no – or scarcely any – distinction between finite and renewable resources, between the products of technology and the products of nature, between global and regional markets. So what can be done, when the traditional formulae for economic growth and redistribution of wealth no longer work?

Analyses of global economic developments since the Industrial Revolution have often created the impression that these developments have come about through the practical application of economic theory. The existence of opposing models at every turn, however, begs the question: why do certain models affect the course of events while others do not? The theories admitted to the canon are usually those which merely describe the events of the time, and which thus serve to legitimize and reinforce those developments. We must be careful to accord theories no more and no less authority than they merit.

World-famous thinkers have either prepared the way for or gone along with systematic ignorance of the fundamental environmental issues of industrial development.[20] There was Francis Bacon (1561–1626), for example, who at the beginning of the 17th century evolved an understanding of nature which sought to deconstruct it into individual fields of experimental inquiry, thereby wrenching them from their natural context. In his utopian novel *New Atlantis*, he prophesied that the natural sciences would pave the way for a technically perfect future state.[21] Or there was Isaac Newton (1642–1727), who saw the natural world as a composite of individual elements, each of which should be studied in isolation in order to be able to better exploit its properties. Or again there was René Descartes (1596–1650), whose principle of systematic doubt helped found modern empirical science – but who also gave rise to the linear thinking that allows no scope for consideration of ecological cycles. The succession continues with Adam Smith (1723–1790), the great exponent of economic individualism and the free market; David Ricardo (1772–1823) and his theory of comparative advantage, which even today underpins proposals to reduce national production costs for the sake of enhanced economic competitiveness; and John Stuart Mill (1806–1873), and the principle he famously espoused, that 'anything that restricts competition is to the bad; everything that promotes it is to the good' and his resultant focus on immediate benefits. And finally, there was Karl Marx (1818–1883), who, while he did document the fundamental conflict between the logic of industrial productive power and the associated relationships of production, neglected to consider the destructive effects of this process on the natural environment.[22]

The object of this book is not to debate these various theories, which were actually more sophisticated than modern rhetoric would suggest, and in any case were based on premises which have long since become obsolete. The question at issue here is, rather, why early theories that did take environmental factors into account failed to gain canonical acceptance. To take a concrete example, why were the ideas of the 'physiocrats', which were widespread in the 18th century, sidelined and forgotten?

It was a tenet of the physiocratic school of thought, originating with the French writer François Quesnay (1694–1774) and his 'tableau économique', that only as much should be taken from nature as it was possible to give back.[23] The wider ecological context is thus at the heart of this analysis of economic processes. Agriculture was regarded as the sole source of new wealth, because it was here that real production took place, rather than mere extraction of resources. Only those processes leading to material increases counted as true production, as opposed to conversion processes which actually resulted in material losses. According to this line of argument, if increases in output must be paid for by destroying resource inputs, economic growth is in reality 'negative growth'. Real growth occurs only where goods are produced from solar inputs – that is, from additional new materials.[24] History rolled over the physiocrats because the science of the time was still incapable of recognizing the finite limitations of resource reserves and the environmental costs of manufacturing processes; because the theory was too complex and unwieldy to stand up to the linear expansion of the Industrial Revolution; because the state of technological and scientific knowledge needed to exploit fully solar energy – and which would have allowed the widespread industrial use of solar resources – did not exist; and because there was still no conceivable solar power technology which could have competed with that of the steam-engine. It was not so much the theories of Adam Smith and Karl Marx that occluded those of the physiocrats, as the technology developed by James Watt.

Nevertheless, the physiocrats did formulate the basic principles of what we now call 'sustainable development'. Back in the 19th century, they had already realized that a sustainable resource base was the essential precondition for the 'natural economy' so vital to long-term prosperity – and that this in turn required a regional economic structure based on agriculture. As forlorn as this vision may seem today, the target – and the outcome – of a solar resource economy is a revitalized agricultural primary economy not just restricted to the production of foodstuffs, but which also supplies energy and native materials. With solar energy, solar materials and appropriate associated technologies, and with globally networked but locally accessible ICT, this kind of decentralized economic development is now within reach. This implies a new global division of labour in which the variety of geographical environments opens up a whole range of new and different opportunities for sustainable production.

Re-establishing the natural circular flow of resources is the key to sustainable and environmentally responsible development in the long term. That is far from saying – as is often assumed – that we must turn back the clocks. On the contrary, it means mobilizing the necessary technologies in constructing appropriate economic frameworks. Nevertheless, in order to follow this path, the global economy will have to free itself of its dependency on fossil fuels and fossil resources, and divest itself of the associated infrastructure and business processes. My fourth proposition is: *an economy based on solar energy and solar resources will make it possible to re-establish the links between the development of the economy as a whole and environmental cycles, stable regional business structures, cultures and democratic institutions, links which are essential if the future security of human society is to be guaranteed.* The resulting structures will be a radical departure from the industrial and post-industrial eras. The primary sector, now all but written off, will become the driving force for the economy of the future. The result of the shift away from the existing non-solar energy and resource base will be an agricultural renaissance, enriched by the possibilities of new power-generation technologies and the breadth and depth of scientific knowledge.

This historical process, which is not reflected in any conventional economic forecasts, will not suddenly revolutionize all existing structures. Rather, it will unfold like the process kick-started by the Industrial Revolution, moving at a different pace in different countries and different continents. Those who still think only in the short term must continue to bow to the laws of fossil-fuelled industrialization. Anybody, however, who seeks to shape a new and different future must not let him or herself be mesmerized by the way things are now. He or she must realize what must be – and how it can come about. And he or she must have the long-term vision to develop decisive initiatives that will help set the ball rolling, until the process becomes self-sustaining. But with fossil reserves rapidly nearing exhaustion, and as the threats to the global ecosystem will be felt well before the oil, coal and gas eventually run out, this transition will have to be accomplished far faster than was the Industrial Revolution in its time. Modern technology is what will make this acceleration possible.

From the political to the economic
solar manifesto

My earlier book, *A Solar Manifesto: The Need for a Total Solar Energy Supply – and How to Achieve it*, published in 1994,[25] dealt with strategies for replacing fossil fuel and nuclear power with renewable energy. It describes how – despite the prejudices of the energy sciences and the energy industry – it is possible to meet all humanity's energy needs from renewable sources. It diagnoses the current energy supply networks as the 'leukaemia of the body politic', which can be cured with renewable energy sources. I call for what was a central part of Carl Amery's 'message for the millennium'[26] to be the first political priority for the new century – an 'Agenda 1'. I identify sources of resistance to this fundamental reappraisal, in particular the opposition of the established energy industry. It brands the decades-long neglect of this key issue by governments who have treated it, at best, as a side issue for energy policy and research, as the 'missed opportunity of the century'. Against the 'economy of death' created by the current energy supply network it proposes a whole range of possible policies to expand the market for renewable energy, from the municipal to the international level, in order to enable an 'economy of survival'.

The *Solar Manifesto* helped bring renewable energy to the attention of politicians, businesses and the public. The EU Commission White Paper 'Energy for the Future: renewable sources of energy' of November 1997 has also since affirmed that 'a comprehensive strategy has become essential' requiring 'across-the-board initiatives encompassing a wide range of policies: energy, environment, employment, taxation, competition, research, technological development and demonstration, agriculture, regional and external relations policies'.[27] Even the World Bank has since recognized the primary importance of solar resources for developing countries, and there have been countless declarations from governments and UN agencies which may lead to concrete policy actions. There are already many new enterprises producing solar technology products, a commercial market is beginning to develop, and ever more local authorities and voluntary groups are becoming active. Most importantly, however, the moment people become aware of the potential for lasting, benign energy supplies, they are immediately sold on the idea and give it their support. Such have been my impressions and observations at innumerable discussions and talks I have attended or given across all sectors of society.

A growing number of scientific studies have shown that earlier conclusions, to the effect that renewable energy could only ever make a small contribution to total energy supplies, are unduly pessimistic. It would seem only a matter of time before the great practical breakthroughs start coming on-stream. One encouraging sign, for instance, is the rapid expansion of wind power in Denmark and Germany in only a few years, or the electrification of a growing number of villages in developing countries using solar power. Numerous new solar technology firms have been founded, and some of the established energy providers are beginning to enter the field. Despite the continuing neglect of renewable energy by publicly funded research and development programmes, new technological developments are now debunking the pessimistic prognoses of past decades for the potential of solar resources. There have also been political declarations aplenty. Does this mean that things are now moving the right way – towards a new dawn in the global energy industry?

There are three main reasons why I am now expanding the political *Solar Manifesto* to cover the economic dimension as well:

1   Recent positive developments must not blind us to trends to the contrary: the introduction of renewable energy cannot keep up with the growth in global demand. The proportionate growth in the use of renewable and fossil energy sources favours the latter. Moreover, the influence of the fossil energy industry on energy supply networks is not declining, but growing by the day. The trend towards increasing international industrial concentration is accelerating, as recent spectacular mergers in the energy industry have shown. The deregulation of the energy industry is only the first stage in the formation of transnational energy corporations. The focus of political and public interest is no longer on green taxes designed to raise the price of fossil

fuel-derived energy, but on achieving the lowest possible energy prices for the sake of global competitiveness. Governments are actively promoting this development, although the public must be aware that low prices for fossil fuel energy exacerbate the global environmental crisis and hamper the introduction of renewable energy.

'They do not what they know' – these words from Robert Jungk, with which I concluded the *Solar Manifesto*, must now be sharpened. Those responsible are doing the opposite of what they should be doing. Global economic policy has become obsessed with obtaining low fossil fuel energy prices through market stimulation, because this – allegedly – is what global economic competition dictates. For this reason, renewable energy sources are in acute danger of facing major setbacks, just as they have begun to establish themselves and could become more established still. The danger is that this self-induced 'solar eclipse' could last long enough to extinguish the future prospects of the 21st century before it has even really begun.

If this is to be prevented, it is obviously essential to engage with the fundamental tenet of the fossil fuel industry, that only fossil fuels – at 'globally competitive prices' – can secure the economic existence of companies and economies. How can we break the vicious circle whereby, in fossil fuel-dominated energy markets, the right hand undoes what the left hand has painstakingly achieved – and must still achieve – for renewable resources? To answer this question, the potential of solar resources must be set within the context of global economic trends. It is precisely these trends – which fossil fuel apologists call upon to justify their arguments – that speak in favour of the solar alternative.

2   The ideal of large-scale introduction of renewable energy is, true to say, no longer contested. However, there is one notorious clinching argument which is always raised against the comprehensive and thoroughgoing realization of this ideal: conventional energy sources are assumed to have an economic advantage, whereas renewable energy sources are denounced as a burden that can be borne only in small doses. Even in the conferences of the World Climate

Convention, the discussion revolves almost exclusively around the avoidance and fair division of burdens. The fact that such conferences are held, and that state governments make their own initiatives contingent on prior international agreements, serves to illustrate the current absurd approach – keeping the conventional energy supply structures for reasons of economic security, despite the fact that it is precisely these structures which jeopardize the security of all. If, however, the transition to solar energy were to be seen for the unique opportunity that it is, then these inter-governmental conferences would be superfluous. Each state government would be promoting the change of its own accord. Would anybody have thought it necessary to make the introduction of information technologies contingent on the decisions of an international convention? Anyone making such a suggestion would have been an object of ridicule.

The question of whether this key assumption – the economic advantage attributed to conventional sources of energy – actually holds water is hardly ever asked. There is no truly objective and universally applicable way to assess whether something is economically viable – it always depends on the requirements and conditions, on the identity of the people for whom it is to be viable, at what direct and indirect prices and at whose expense economic viability is to be achieved. The current assumption that fossil fuels are inherently more economical, however, is based on an incomplete analysis of the atomic/fossil fuel energy complex using calculations that cannot be applied to solar energy. This is the basis of my fifth proposition: *an examination of the entire supply chain for fossil fuel energy demonstrates that its claim to be more economical is a myth. In theory, renewable energy sources have an economic advantage because of their much shorter supply chains. This can be exploited if the atomic and fossil fuel energy suppliers are divested of their numerous state privileges, and technical development and market introduction strategies for renewable energy are refocused on this unique economic advantage. Only then can the switch to renewable energy acquire its own unstoppable momentum.*

Technologies, power sources and materials, as the history of the post-Industrial Revolution period shows, all have their own economic logic, which the manufacturers of the day follow until an optimum use of the relevant development has been achieved. However, this process is usually understood solely in terms of cost reductions through technological development and increased productivity; structural considerations are not addressed. The same is true of the debate on renewable energy. While the fundamental difference between renewable and conventional energy sources with regard to the environmental consequences is recognized, their relative economic viability is assessed solely on the cost comparison between isolated generation technologies, and not on the basis of what is economically relevant prior to or following the exploitation of these technologies.

If we examine the supply chain necessary to exploit solar energy, then we can see that – to expand on my fifth proposition – *renewable energy sources can be harnessed in a more efficient, more user-friendly and thus more economical fashion than would ever be possible using conventional energy sources. This, however, can only be achieved if renewable energy sources are employed quite independently of conventional energy supply chains.* The same applies to the use of solar as opposed to fossil and mineral resources. Technological breakthroughs are still required, of course, before these benefits can be realized in full. These, however, can be relatively precisely specified and achieved because they depend not on unpredictable coincidences, but on the known possibilities of physical laws. Power storage techniques in particular – hitherto neglected in solar energy research and development – will be needed. If manageable and cost-effective power storage technologies are available, then the revolutionary transition from the fossil fuel to the solar economy will be unstoppable.

3   The debate on the economic potential of solar power has hitherto accorded only a secondary role to solar resources. Even some ecologists are unaware that solar resources may provide a raw materials base that goes far beyond isolated uses.[28] Biomass is predominantly viewed more as an energy

source than as a raw material in its own right. Many commentators fear that the aims of ecologically sound agriculture will be defeated if industry seeks to exploit this potential on a large scale.

Although solar resources have the same basic properties as solar energy sources, these do not apply to every conceivable scenario for their exploitation. They could, for example, fall into the monopolist hands of global corporations. Seed and land are intermediaries between the sun and the materials produced via photosynthetic processes, and agrochemical corporations will seek to control this raw materials market as they already do in the case of food products. If solar resources meet the same fate, then the opportunities they provide would be under-utilized or even missed entirely, both from the environmental and the socioeconomic perspective. The key to the exploitation of solar resources is therefore to recognize and realize the opportunities available for promoting and preserving economically viable local agricultural business structures.

The most sensitive question humanity faces is whether the global economy produces enough to go around. If our economy continues to be based on limited, polluting resources and ever more concentrated global business structures, then there will not be enough for all. The more obvious this becomes, the more likely it becomes that in the absence of a clear alternative, the ideal of equal human rights will be revoked. This process is covertly already well under way. Carl Amery identifies the nub of the issue: if there is not enough to go around, then the Nazi doctrine of 'national selection' will not remain an isolated historical episode. Instead, the distinction between the privileged and the disenfranchised, between those seen as superior and those seen as inferior, will be maintained into the 21st century.[29] We face the threat of new genocides in new wars for *Lebensraum*, and of 'ecocides', brought about by humans and guaranteed by the law of the market. But when nature strikes back, she has no regard for privilege; her selection is indiscriminate. She is just and yet unjust, for her vengeance also meets those who have not provoked her. Yet she will tolerate a compromise, which we ourselves must enact. The answer is not 'environmental protection', which merely maintains isolated reserves without arresting the overall destruction, but rather a natural economy which respectfully partakes of the 'Wealth of Nature' (Donald Worster),[30] instead of disfiguring the world with rape and pillage in the pursuit of an imagined 'Wealth of Nations' (Adam Smith).

The goal of universal provision is the social and democratic ideal of the modern age, an ideal which originated with the Industrial Revolution. But the Industrial Revolution's excesses, which have led us to put ourselves above nature, make it impossible to realize this ideal for all people in the long term. To achieve universal provision, it is not necessary to give nature priority over the needs of humans. *What is essential – and this is my sixth proposition – is the primacy of physical laws over the laws of the market. In practical economic terms, this means above all that locally or regionally produced solar energy, foodstuffs and solar resources should be consumed and marketed in preference to otherwise equivalent products.* A society which, with the aid of its political institutions, is unable to reverse the primacy of the market over nature is destined to die. The choice is not between private or public enterprise, between the free market or the planned economy. It is a question of the physical laws that govern private and public enterprise, market and planned economy alike.

Solar resources are products of the primary sector. In view of their fundamental importance in providing for the inhabitants of an economic region, they may not be subordinated to the market or to some macroeconomic plan. This is the essential conclusion that follows from the sham existence of the fossil-fuelled global economy. By switching to a solar resource basis, we can end this sham, and ensure that there will be enough to go around.

My seventh proposition is that *only a solar global economy can satisfy the material needs of all mankind and grant us the freedom to re-establish our social and democratic ideals.* This solar global economy will consist of a global market for the products of technology alongside innumerable linked regional commodity markets, whose economic basis cannot be usurped. An unattainable utopia? On the contrary, it is the predominant belief that a stable economic future can be achieved through the famous 'invisible hand of the market' which is utopian. According to Adam Smith's theory, the market directs individual behaviour such that people's unconscious interaction works to the general good. Although many times disproved, foiled by the market participants themselves, this theory has become an axiom, an indisputable absolute truth, of which the collapse of the socialist planned economy seemed to be the final proof. But the directly perceptible hand in each case is more often greedy than helpful, more often taking than giving, more often combative than sympathetic. Being invisible, the hand of the market can steal and exploit without being recognized. The result is not harmony, but tension, division and disruption.

My proposal, to rely above all on the *visible hand of the sun*, on desirable and direct benefits, is more precise, more comprehensive, more manageable, more comprehensible, more accessible, more appropriate to people's needs, and more realistic. It is also free of danger and definitely less utopian.

The propositions in summary

1 Global civilization can only escape the life-threatening fossil fuel resource trap if every effort is made to bring about an immediate transition to renewable and environmentally sustainable resources and thereby end the dependence on fossil fuels.

2 Making the groundbreaking transition to an economy based on solar energy and solar resources will do more to safeguard our common future than any other economic development since the Industrial Revolution.

3 Economic globalization can only be made environmentally sustainable through the targeted replacement of fossil fuels by solar energy sources. This is the only way to rein in the destructive imperative of the fossil economy and call a halt to the creeping homogenization of economic structures and cultures. It is the only way to make economic development diverse, sustainable and of lasting benefit to both individuals and society.

4   An economy based on solar energy and solar resources will
    make it possible to re-establish the links between the devel-
    opment of the economy as a whole and environmental
    cycles, stable regional business structures, cultures and
    democratic institutions, links which are essential if the
    future security of human society is to be guaranteed.

5   An examination of the entire supply chain for fossil fuel
    energy demonstrates that its claim to be more economical
    is a myth. In theory, renewable energy sources have an
    economic advantage because of their much shorter supply
    chains. This can be exploited if the atomic and fossil fuel
    energy suppliers are divested of their numerous state privi-
    leges, and technical development and market introduction
    strategies for renewable energy are refocused on this unique
    economic advantage. Solar resources can be harnessed in a
    more efficient, user-friendly and thus more productive way
    than would ever be possible with conventional energy.

6   The immutable laws of physics must have primacy over the
    mutable laws of the market in our economic order. It follows
    from this that locally or regionally produced solar energy,
    foodstuffs and solar resources should be consumed and
    marketed in preference to otherwise equivalent products.

7   Only a solar global economy can satisfy the material needs
    of all mankind and grant us the freedom to guarantee truly
    universal and equal human rights and to safeguard the
    world's cultural diversity. What is in principle impossible
    with the 'invisible hand of the market' alone can be achieved
    with the visible hand of the sun.

PART I

CAPTIVITY OR LIBERATION: FOSSIL FUEL AND SOLAR SUPPLY CHAINS COMPARED

The importance of our resource base to our economic wellbeing is hard to overestimate. Yet comparative studies of resource productivity attract more misconception, half-truth, blindness and simple bloody-mindedness that almost anything else. Even where the individual characteristics of different resources are known – locations of deposits, necessary extraction and refining techniques, applications, market participants, prices, achievable efficiencies, quantities and consequences of resultant emissions – these are almost always discussed in a disconnected and fragmented way. Fossil fuels are assumed to offer lower prices and greater potential, whereas solar energy is thought to have the edge solely in terms of reduced environmental impact. Very few people are aware that different resource types necessitate different economic structures, and promote different developmental trends. This goes above all for the majority of established experts, who have had all notions of a holistic approach systematically beaten out of them by the culture of technical specialization in the science and business worlds.

In order to comprehend the scope of the resource question, we must conduct a systematic analysis and evaluation of the differing supply chains, from the various primary resources through to the end-users. The logic of business and technol-

ogy that governs economic behaviour can only be fully understood from the perspective of the supply chain and its
dependency relations. This supply chain analysis is far more
important than individual power stations or commodities. A
piecemeal approach that contents itself with finding solar
replacements for individual power plants or commodities
cannot hope to free the world from the web of supply chain
links with which the fossil fuel and resource industry has
girdled the planet.

The fundamental economic reality of fossil fuels is that
they are found in only a relatively small number of locations
across the globe, yet are consumed everywhere. The economic
reality of solar resources, by contrast, is that they are available,
to varying degrees, all over the world. Fossil fuel and solar
resource use are thus poles apart – not just because of the
environmental effects, but also because of the fundamentally
different economic logic and the differing political, social and
cultural consequences. These differences must be acknowledged
if the full spectrum of opportunity for solar resources is to be
exploited.

1

Ensnared by fossil supply chains

MANY AUTHORS HAVE charted the course of civilization through the development of its energy systems — for instance, Debeir, Deleage and Hémery in their history of energy systems,[1] Smil in *Energy in World History*,[2] and Sieferle in his work on the history of mankind and the environment.[3] They all mount a convincing attempt to trace out how the resources used at each point in history shaped the economic, social and cultural trends of the time. There is no need to reiterate the detailed paths taken by individual economies here. Cultural differences notwithstanding, the process is more or less the same whenever and wherever it takes place. The aim here is to describe how the world is fettered by the supply chains of a finite, exhaustible resource base, chains which are dragging humanity inexorably into the abyss.

Long supply chains due to limited resources: the logic of globalization

Globalization came about as the industrialized countries began to exploit global fossil fuel and mineral reserves in pursuit of greater economic efficiency. Sparse reserves begat long supply chains with equally extensive consequences. The modern demand for fuel and mineral resources is the real driving force behind globalization. Equally, ever increasing demand in all its various forms for fossil fuel and mineral resources is the only compelling reason why autarky is no longer an achievable goal for market economies. The drive to annex global fossil fuel and mineral reserves has produced an ineluctable pressure to global-ize, whereas locations for the manufacture of finished goods or provision of services may be shaped on a regional basis. Finished goods and services are always changing, but demand for resources – and energy resources in particular – is a constant presence which can be reduced only by economizing on use. The 'Silk Road' to China, the discovery of the Americas and of Australia, the opening up of southern and central Africa and forced colonialization all paved the way for increasingly global markets. New opportunities were provided by infrastruc-tural improvements such as expanded transport links, better communications technology and the growth of international capital markets. However, it is only the demand for fossil fuel energy and resources, together with their associated industries, that made a lasting impact on the structure of global society. As a result, not only did the industrialized countries become dependent on the exporters of energy and mineral resources, but the exporting and consuming countries alike became depen-dent on the global energy and mineral resource industries.

The crude oil supply chain

The first link in the supply chain is restricted to those few countries possessing the oil reserves to support a domestic extraction industry. The notable reserves, which have attracted enormous capital expenditure, are located in the USA and Mexico, Argentina and Venezuela, in the North Sea, in the Caucasus, in Nigeria and Somalia, in China and Indonesia and above all on the Arabian peninsula. Extraction has become a high-tech and thus highly capital-intensive industry, especially in the case of secondary oil recovery, in which the last drops are wrung out of an oil field. Secondary extraction techniques range from flooding with water, polymers, carbon dioxide (CO_2) or corrosive solutions through to water and gas injec-tion. All these procedures may result in serious environmental damage long before the oil leaves the well. The extracted oil is then transported, often over thousands of miles, via energy-hungry and accident-prone pipelines and pumping stations, in supertankers or tanker-trains, to the refineries of the industrialized countries. The refineries – the third link in the chain – crack the oil using fractional distillation, converting it into fuels and feedstocks for the chemical industry. The refining process causes even more environmental problems than extraction: emissions of hydrocarbons, sulphur, nitrogen and carbon monoxides, and liquid and solid wastes. The consequence is the fourth link in the chain, which is waste disposal. The fifth link is the storage of refined products, and the sixth is the shipping of fuels to garages and of other products to their onward destinations. Fuel combustion in engines, furnaces or power stations and feedstock consumption by chemical plants form the seventh link.

The natural gas supply chain

Natural gas reserves are also found in only a few countries and regions, principally Russia, the Caspian Sea region, Iran and Algeria. Gas extraction is not a simple process either, as the gas must be both purified and condensed before it can be transported. By-products of these processes include sulphur and fertilizers. Depending on the ultimate end-use and means of transport employed, the gas may be liquified, a process that uses temperatures of $-162°$ Celsius to achieve a 600-fold reduction in volume, and which requires enormous quantities of energy. For technical reasons, this refining process often takes places at the point of extraction. Liquid gas mixtures are also produced for the petrochemicals industry and to provide industrial process energy. The third link in the chain is the transport through pipelines and their compressor stations to storage tanks, often over thousands of miles (for example, from the Caucasus to central Europe, or by supertanker from Algeria to the USA). Transport and storage tanks for liquid gas are costly to construct: they have to be very well insulated to maintain the low temperatures, and require energy-intensive cooling systems. The fifth link in the chain is the distribution through regional pipelines or in gas tanks to the end-users – private households, power companies or the manufacturing industry (to fuel high-temperature processes) – who then (sixth link) burn the gas in power stations, boilers or combustion engines.

The coal supply chain

Today's major coal exporters are Australia, the USA, South Africa, Canada, Russia and Poland. The distribution of reserves is relatively wide, and some countries which consume comparatively large quantities of coal are able to meet their demand entirely from domestic sources. The country which is by far the most heavily dependent on coal is Japan, where more than a quarter of the total world output is consumed. Within Europe, the major importers of coal are the Netherlands, Denmark, France, Italy and Spain.

Due to the wide variation in the type and quality of coal deposits, coal extraction is a highly complex process. Coal deposits differ greatly in their water and sulphur content and in the degree to which the deposit is mixed in with other material, and the extraction techniques of open-cast and shaft mining are equally diverse. In the second link in the chain, following extraction, the coal is refined to suit differing needs. The raw coal is first graded, and foreign bodies are removed. Then come crushing and homogenization, crude and fine sorting, and dehydration. Finally, the coal is turned into either briquettes for small-scale combustion, power station coal, or coke for use in blast furnaces. The refining process is particularly costly in the case of brown coal (lignite): the coal must be dried, broken, sieved, ground and dried a second time in order to reduce the water content from over 50 per cent down to 10–20 per cent. Only when this stage is complete can it be turned into briquettes, lignite dust for industrial furnaces or coke. The third link comprises waste disposal processes: sludge thickening, mineral enrichment (flotation), flushing and filtering. Large quantities of energy are consumed during refining and waste disposal, and extensive water pollution results. In the fourth link in the chain, the refined coal is shipped to the fifth link, the power stations or retail consumers. As with oil and gas, the coal industry has also seen a consistent upwards trend in its transport distances.

The nuclear supply chain

The most complex supply chain is that of the atomic energy industry. The first link, extraction, is complicated by the danger of radiation. In the second stage, the uranium ore is transported from countries such as Australia or Canada to refining plants where the ore is turned into uranium oxide. This so-called 'yellowcake' is the third link in the chain. In the fourth and fifth stages, the yellowcake is transported to processing plants for the production of uranium hexafluoride. In the sixth link, the uranium hexafluoride is shipped to a uranium enrichment plant, where production of the actual fuel rods forms the seventh link in the chain. The fuel rods are then shipped (eighth link) to the power station. Each individual step involved in the extraction and processing of uranium ore is accompanied by the intensive use of technology, high energy use, considerable environmental damage and huge risks.

The following statistics illustrate just how dramatically the industrialized countries' dependency on imports has increased: between 1975 and 1994, German imports of fossil fuels increased from 115 to 160 million tonnes; Japanese imports from 475 to 555 million tonnes; and US imports from 1.77 to 2.2 billion tonnes.[4] In its White Paper on renewable energy, the European Union (EU) Commission calculated that Europe's dependency on fossil fuels could grow from 50 to 70 per cent by 2020.[5] Germany already meets 70 per cent of domestic energy demand from imports; for crude oil, the level is already almost 100 per cent.

The lengthening supply chains in the electricity generation industry

On top of the supply chain links already discussed (seven for oil, six for gas, five for coal and up to nine for nuclear fuel), there are also the waste disposal costs and the distribution grids of the electricity suppliers: high-voltage transmission to regional substations where the current is transformed to medium voltage, followed by transmission to local substations and subsequent low-voltage transmission to the end-users. The last link in the chain is the use of electricity to power lights, heating, motors or other electromechanical or electrochemical processes. The total length of the resulting supply chain, from extraction to end-use, is:

- at least ten links for coal-fired power stations (one link fewer for gas-fired power stations, because gas combustion leaves no residues to be disposed of); and
- at least 14 links for nuclear power stations (in the case of reprocessed fuel, at least seventeen).

These figures take no account of the supply chains involved in the construction of extraction facilities, pipelines, tankers and freighters, power stations and cabling, nor of the need to deal with land and water pollution, nor indeed of the damage to human health and to the climate caused by individual links in the chain.

Resource and mineral supply chains

Industrial resources comprise firstly the various types of quarried stone and rock, sand and mineral salts which are the oldest and most abundant of the non-renewable resources, and which can be extracted at numerous locations across all continents; secondly, mineral ores in the form of compact deposits, which have shaped the world since the dawn of the industrial age; and finally, the hydrocarbons extracted from coal, gas and above all from crude oil.

In metals extraction, the ore must first be separated from the surrounding rock, which produces large quantities of spoil. The extracted ore is then processed – the second link in the chain – to separate the crude ore from the useless and harmful components, and press, sieve and break it up ready to feed to the blast furnaces. Processing facilities are usually located near the mine. Rock with low concentrations of ore has to be enriched, which involves milling in order to be able to admix other minerals. Total world output from iron mining is more than 800 million tonnes annually. Both of these first two links in the chain are highly energy-intensive. The third step is to ship the processed ore to steelworks across the globe, using freighters which – again employing large quantities of energy – traverse routes stretching as much as 12,000 miles from the two major exporting countries, Australia and Brazil, to Europe and the USA.

Before they can be processed further, most mineral ores require additional refining (fourth link) to remove amalgamated substances, and extract pure metals and other materials that can be used in the production of synthetics, composites, alloys or minerals for use in fertilizers and pharmaceuticals. This process, too, consumes copious quantities of energy. The ore is then shipped to the smelting plants where the actual metals are produced, and the finished metals are delivered to customers' premises. The supply chain for metals is thus as a rule at least six links long. Other mineral raw materials such as gravel and sand, potassium or salts also feed into other supply chains, primarily in the chemicals industry.

Table 1.1 *Geographical concentration of mineral reserves*

Commodity	Proportion of known global reserves possessed by the three largest exporters	
Platinum	99.5%	South Africa, Russia, Canada
Chromium	96.9%	South Africa, Zimbabwe, Russia
Vanadium	94.9%	Russia, South Africa, Chile
Manganese	90.5%	South Africa, Russia, Australia
Asbestos	81.3%	Canada, Russia, South Africa
Molybdenum	74.3%	USA, Chile, Canada
Tantalum	72.7%	Zaïre, Nigeria, Russia
Tungsten	69.6%	China, Canada, Russia
Mercury	65.2%	Spain, Russia, Yugoslavia
Aluminium	63.8%	Guinea, Australia, Brazil
Cobalt	63.0%	Zaïre, New Caledonia, Russia
Iron	59.4%	Russia, Brazil, Canada
Titanium	59.0%	Brazil, Canada, India
Silver ·	54.9%	Russia, USA, Mexico
Nickel	54.5%	New Caledonia, Canada, Russia
Tin	50.2%	Indonesia, China, Thailand
Bismuth	47.9%	Australia, Bolivia, USA
Lead	47.8%	USA, Australia, Russia
Zinc	45.8%	Canada, USA, Australia
Copper	44.9%	USA, Chile, Russia

Source: Bilardo/Mureddu: *Energy, Raw Materials for Industry*[6]

Most European industrialized countries are now 100 per cent dependent on imports of metal ores.[7] This dependency continues to grow, because most industrialized countries have already exhausted any deposits they once had in the course of their industrial development. In 1980, for example, the then members of the European Economic Community (EEC) produced 5 per cent of their antimony ore, 2 per cent of their manganese and mercury, 4 per cent of their copper and 12 per cent of their nickel. Domestic extraction of iron, chromium, germanium, cobalt, molybdenum, niobium, platinum, titanium and tungsten had already ceased.[8] Even the USA, which by virtue of its sheer size is rich in mineral deposits, is reliant on imports for many metals: 100 per cent for titanium, niobium, tin, germanium and platinum, 98 per cent for manganese, 96 per cent for tantalum, 90 per cent for chromium and cobalt and 70 per cent for nickel. The figures come from a study performed by the Pentagon's Energy and Defense project.[9] Although it is possible in many cases to substitute one metal for another, the limited availability of local deposits means that this does not fundamentally reduce the extent of the dependency.

Fossil resource supply chains and industrial concentration: market destruction through market mechanisms

The tendency for only a few or only one large firm to survive in a marketplace initially comprising many competing firms is simultaneously regarded both as the normal course of development in a market economy, and as presenting a danger to it. Mergers and acquisitions are justified in terms of economies of scale, and latterly also by reference to 'synergies', that is, the ability of firms to complement and expand on each other's specialized skills and technologies. All market economies have made legal provisions for interventions to prevent the formation of cartels and monopoly abuses. Such measures may postpone industrial concentration, but they cannot stop it. As businesses become multi- and transnationally organized, a process facilitated by the General Agreement on Tariffs and Trade (GATT), the WTO and continental market regimes like

that of the EU, which is designed to prevent economic protectionism, political measures to combat cartels are losing their
effectiveness. The political goals of international free trade on
the one hand and the prevention of cartels on the other are
thus increasingly at cross-purposes. Measures to combat
protectionism have played an obvious role in facilitating
mergers and acquisitions. International trade agreements accord
transnational corporations a de facto privileged position and
promote giant mergers. The result is the transformation of the
global marketplace into a market for the few, with global market
forces being cited as justification for the resultant erosion of
competition.

Unlike all other industries, in the minerals and energy
industries the pressure for increased industrial concentration
derives directly from their business models. It is by no means
simply a product of the pursuit of increased productivity
through greater business scale, but the result of extended global
resource supply chains. By driving globalization and industrial
concentration in the energy and commodities business, global
resource supply chains have also given a decisive impetus to
industrial concentration processes in the economy at large. If
it were not for the highly concentrated availability of wholesale energy supplies, merger activity would probably have
followed a lower-key, more differentiated course.

The high cost of prospecting alone, requiring countless
geological surveys and test drillings, can be borne only by
capital-rich firms. Only wholesale investors with guaranteed
long-term sales can afford investments with such lengthy
amortization periods. The same applies to the use of modern
extraction techniques, the construction of pipelines and the
provision of large-scale freight capacity. For oil, coal and ore
shipments, freighters with up to 800,000 tonnes carrying
capacity are used; for gas shipments, capacities run to 200,000
cubic metres of liquid gas. Such large deliveries necessitate
large refineries and high storage capacity, which means centralized plant and high-volume storage. Processed materials and
energy are shipped onwards to equally large power and smelting plants, for the same reasons of economic scale. The sheer
weight of these concentrated material flows leads to the forma-

tion of strategic alliances between large-scale resource suppliers and the operators of large-scale industrial processing and energy-generation capacity. There is pressure to manage the whole supply chain in-house, or at least to control it. The oil giants led the way in this regard, bringing the whole chain from exploration through to garages under one roof. Conglomerates uniting coal and ore-mining firms with electricity generators and smelting plants also crystallized early, for the same reason – on the national level, for as long the domestic reserves of the industrialized countries lasted, then on the international level. Since the beginning of the 20th century, during which oil replaced coal as the main source of energy, internationalization due to the geographically restricted distribution of oil reserves has been a foregone conclusion. The oil giants – the infamous 'seven sisters' – became the first 'global players', and thus the exemplar of 20th-century business.[10]

Exporting countries have frequently tried to strengthen their position through the use of state-owned companies to exert control over their reserves, and preferably downstream operations as well. One such attempt is the Organization of Petroleum Exporting Countries (OPEC) cartel; another is UNCTAD (the United Nations Conference on Trade and Development), which every four years tries to force through fair prices for resources – almost entirely without success, because the importing countries, and the resource giants in particular, can play the exporters off against each other almost at will.[11] The resource conglomerates, which control transport capacity, processing plants and power stations, and the energy and materials markets in the consuming countries, have the upper hand. They have long since been the financial backers, partners, shareholders or owners of extraction companies in the exporting countries.[12] They act, in effect, as new colonial powers – but without accepting any political responsibility.

In the sphere of electricity generation, which began in the industrialized countries with water and steam power, this level of concentration was harder to achieve than in the pure oil, gas and coal supply chains. This is because retail distribution depends on the local electricity grid, which could not so easily be taken over without political help and connivance. The technical and economic advantage that derives from the control of large-scale hydropower, much the easiest and cheapest source of additional and reserve capacity, was a trump card that the power companies were able to play to their advantage. They constructed the national grids, which they used to monopolize supply, freeze out local electricity producers, and now finally to take over ever more municipal distribution grids. As wholesale purchasers of fossil fuels, they were able to demand lower prices and subsequently to vary their prices locally in order to price municipal suppliers out of the market. They also froze out local producers – such as the operators of small-scale hydro plants and wind turbines, large numbers of which were in operation in the 1930s in places like the USA, Denmark and Germany – even where this could not be justified on cost grounds. Despite the fact that many independent operators of small-scale hydro plants and windfarms could produce electricity cheaply, the grid monopolists either refused to purchase their current, or offered prices that were insufficient to cover the producers' costs. Local and municipal power plants, after all, presented an obstacle to the erection of a comprehensive monopoly on electricity supply.

The power companies' greatest trade advantages, however, were and remain the political privileges that they were granted as, with the growth of the electrical goods industry, the importance of electricity generation in social and economic strategy came to be recognized, and as demand for electricity grew. The power companies positioned themselves as guarantors of a stable and uniform electricity supply and, in response, the statutes regulating the power industry were tailored to suit their needs. In other words, governments actively promoted increased concentration in the industry. Concentration became the leitmotiv of capitalist heavy industry, which sought to base its activities on carefully planned guaranteed deliveries. It became the social democratic ideal for the state to come, as described by Ballod-Atlanticus in his paean to large-scale power stations.[13] It became a communist dictum, as embodied by Lenin's famous statement that communism was composed of Soviet power and electrification. Concentration also became the basis for military strategy, which was the reason why the

German Energy Act of 1935 actively promoted the centralization of the energy industry.[14] In short, concentration became the unifying factor in capitalism, fascism, communism and social democracy, and the principal goal of industrial societies of all natures.[15] The nationalization of the French energy industry with the foundation of Électricité de France in 1946, the nationalization of the Italian energy industry under the umbrella of ENEL in 1962, and the foundation of the Austrian national grid company are all artefacts of this process.[16]

Where governments did not themselves drive the concentration process, either through nationalization or by means of energy legislation, the energy industry took matters into its own hands. Its representatives bribed municipal politicians to close municipal plants or sell municipal distribution grids, as described by Lutz Mez in his work on the expansion of the RWE group.[17] It blackmailed local authorities and used grid access restrictions and sabotage to put pressure on local producers, as reported by Berman and O'Connor with reference to many examples in the USA. It was developments like this that roused the ire of Tom Johnson, who was Mayor of Cleveland at the turn of the 20th century: 'I believe in municipal ownership of monopolies. If you don't own them, they'll own you. They'll destroy your politics, corrupt your institutions and ultimately deprive you of your freedom.'[18]

The liberalization of electricity markets across the world and the dismantling of regional and national monopolies has led many to believe that the concentration of market power in the electricity industry has now run its course. Nothing could be further from the truth. The industrial concentration process has in fact received a wholly new impetus. The trans-European gas and electricity grid is being expanded with political backing from the EU Commission and subsidies from the Trans-European Networks programme. The legislative framework for the common market in electricity, which became law in 1997, has given a new significance to the West European grid (UCPTE), and plans are afoot to link it with the common energy grid of Poland, Hungary and the Czech and Slovak Republics (ENTREL), the Baltic ring and the Russian grid (EES).[19] The former public energy utilities are now being sold off, and the resultant wave of mergers is creating the first transnationally organized energy producers. Tariffs have fallen globally with the inclusion of energy in the WTO regime. On a European level, the new European Energy Charter is designed to extend the protection of international law to energy investments in other countries (meaning investments in extraction facilities on oil, coal and gas fields). All these developments are helping to strengthen supply chains and concentrate energy flows, with the aim of flooding the markets with more and cheaper energy.

Despite lip-service commitments to sustainable development, few seem disturbed by the fact that the processes outlined above are accelerating the exhaustion of available reserves. Environmental objectives continue to be stymied, irrespective of national and international decisions on climate change. On the contrary, the loss of regionally protected monopolies is being compensated for through increased market concentration and internationalization, with heightened and accelerated flows – with the result that the opportunities provided by the opening up of the energy markets are negated by the heightened market power of the grid operators. This is a dangerous development. As absurd as it may sound, market mechanisms are destroying the market.

The spider in the web: the growing influence of Big Energy and Big Mining

The energy and minerals industry is highly concentrated, composed for the most part of local monopolies. It has become the focal point for the formation and entrenchment of cross-sectoral industrial cartels that have paralysed the economy in the face of growing environmental challenges. These concentrations of economic power are organic outgrowths of energy and material flows. Like a spider, the fossil resource industry has been spinning its web over more and more sectors of the economy. Each strand of this web is a supply chain, with cross-links composed of other directly connected industries. Yet for almost each and every node in this industrial web, there is an undisputed supply chain logic and a convincing business case.

The web of Big Oil: the oil–petrochemicals complex

Oil refineries do not just produce petrol and diesel. Of the crude oil input, 45.6 per cent is turned into petrol, 20.9 per cent into diesel or heating oil, 9.4 per cent into kerosene and 1.3 per cent into naphtha-based fuels for jet aircraft, 6.8 per cent into residuum, 1.2 per cent into lubricants, 2.9 per cent into petrochemical feedstocks, 3.2 per cent into asphalt, 3.9 per cent into petroleum coke for carbon electrodes, among other uses, and 3.6 per cent into liquid gas.[20] It is possible to vary the proportionate quantities of the various outputs, but only within a limited range. Refineries are thus a focal point where industrial interests find common cause. Each individual interest may be motivated by sound business reasons, but in concert they generate an unhealthy resistance to change.[21] The automobile industry has an interest in maintaining low prices for petrol and diesel products. The aviation industry seeks to secure sufficient supplies of kerosene; the shipping and heating industries likewise seek supplies of diesel and heating oil. Finally, there is the chemicals industry's demand for hydrocarbons for the production of fertilizers and pesticides, and the interest of the refinery operators themselves – usually elements of the chemicals industry – in maintaining demand in the right proportions for all refinery outputs.

A disproportionate change in the demand for one output can displace sales of other outputs. An enduring imbalance in demand results in increased costs. For example, the demand for diesel-driven cars should not grow faster than the demand for petrol-driven ones. If demand for kerosene increases, which is currently the case due to the rapid expansion in air travel, the oil industry is forced to seek additional markets for its other outputs or to flood the market with cheap products. In this system of mutual interdependencies, the optimal balance is achieved when demand grows uniformly across the board. Refineries are the anvil on which alliances between the crude oil, chemicals, automobile, aviation and transport industries are forged. The primary axis is the common interest of the crude oil and chemicals industries. The automobile and aviation industries profit from the availability of cheap fuel, and the fuel producers profit from bulk sales and the production of feedstocks for chemicals, fertilizer and pesticide manufacture. What unites them all is their common interest in stable relative demand, helping one another to achieve a balanced increase in consumption and to head off political interventions that might lead to demand gaps in one market or another.

Much is explained by these considerations: the enduring reluctance of the automobile industry to develop fuel-efficient vehicles, although this would have no direct impact on car sales; the otherwise incomprehensible refusal of the oil companies to market even lubricants from environment-friendly vegetable oils, which would free up crude oil derivatives for other uses; the all but point-blank refusal of companies to develop alternative fuels and engines. The effect on refinery output also explains why even the chemicals industry is vehemently opposed to increases in fuel duty, although at first sight this does not seem to affect them. This analysis also demonstrates that it is a serious political error to seek to impose 'green' taxes only on one oil derivative – fuels for road vehicles – rather than on an upstream link in the chain. Either imports of crude oil must be taxed directly (which could lead to the relocation of refineries), or all products must be taxed equally. By now it should have become absolutely clear that if real progress towards an alternative is to be made, the entire energy supply chain must be taken into account, right down to the ties between individual links. It must also be clear that the radical approach is the only one that promises any success.

The web of Big Gas: the gas–chemicals–oil complex

Functional divisions within the gas-processing industry are similar to those in oil refining. Natural gas has several components: 70–80 per cent is methane, with smaller proportions of ethane, propane and butane, plus nitrogen, hydrogen sulphide, helium, sulphur and water. Apart from gas for the energy industry, gas-processing outputs include liquid gas for the petrochemicals industry and for industrial process energy, as well as the production of chemicals such as acetylene, methanol, chloroform and formaldehyde, and gas feedstocks for the manufacture of a whole range of other synthetic chemicals. All these products have their own storage requirements. For these reasons, the gas industry is heavily interwoven with the chemicals industry. Not by chance is Wintershall, one of the three largest gas importing businesses in Germany, a wholly owned subsidiary of the chemicals giant BASF, which, together with the world's largest gas supplier, the Russian-owned Gazprom, also has joint ownership of two companies involved in gas extraction and pipeline construction. As the oil giants are now also increasingly moving into gas, the webs of the gas and oil industries are becoming ever more closely interlinked. Partly, this is because oil companies are taking a strategic position against the exhaustion of oil reserves, with natural gas providing a replacement revenue stream when the oil runs out. And with the current rate at which new gas-fired power stations are being built, the gas and electricity webs are also becoming ever more closely linked.

The web of Big Mining: the energy–minerals complex

A similar process is underway in industrial processing of mineral resources. Minerals are also processed in refineries in order to separate out amalgams, extract pure metals and manufacture composites and alloys. This processing takes place mostly in the industrialized countries, so that outputs can easily be shipped to customers in the metalworking industries. Developing countries where the ores are mined receive only one tenth of the over $100 billion of investment conducted annually. The refined products find use as metal, as industrial components and as amalgams in the steel, metalworking, electrical and electronics, petrochemicals, paints and dyes and glass industries.[22]

Fear that flows of processed minerals might be restricted, diverted or simply dry up also promote structural conservatism and cartel formation in the minerals industry. The metals-processing industry also works arm-in-arm with the energy industry. The aluminium industry, for example, is opposed to higher energy taxation because energy makes up a high proportion of its cost base.

The web of Big Electricity: the energy–industrial combine

Within the electricity industry, the triumvirate of primary energy suppliers, power station operators and distribution grid companies has to find the right mix of the various types and capacities of energy sources. Nuclear power and lignite are preferred where the load is constant, lignite and black coal for times of medium load, and large-scale hydro is preferred for peak loads on national grids. Interlopers are not welcome – not windfarms, not small-scale hydro, not municipal operators generating from local sources, nor anything that might cause the large power stations to run at less than full capacity. This triumvirate has the backing of the large investment banks. A large power station represents an investment of the order of several billions of dollars; lead times are long, requiring long-term finance, and returns on the investment occur significantly later than with other large-scale investment projects.

Energy-sector investment – of which one third falls on electricity generation and supply – makes up roughly 15–20 per cent of total domestic investment.[23] As the finance for these investments is largely provided by the internationally active investment banks, it is safe to assume that these provide on average 40–50 per cent of all lending in the energy sector. Between 1988 and 1997 alone, investment in electricity supply systems in Germany totalled more than DM126 billion (€64.4 billion; $57.3 billion), with DM44.6 billion (€22.8 billion; $20.3 billion) for generating capacity, and DM61.2 billion (€31.3 billion; $27.8 billion) for distribution infrastructure. Electricity companies undertake the majority of these investment projects, and almost all investment in generating capacity went on fossil fuel plant and related transport and distribution infrastructure.[24] Of the DM11.6 billion (€5.9 billion; $5.3 billion) spent in 1997, nearly DM2 billion (€1 billion; $0.9 billion) went on high-voltage distribution, DM1.6 billion (€0.8 billion; $0.7 billion) on medium-voltage distribution and DM2.1 billion (€1.1 billion; $1 billion) on low-voltage distribution. Annual investment in gas supply exceeds DM5 billion (€2.6 billion; $2.3 billion), of which three quarters is spent on piping. Two thirds is spent on local supply, one third on long-distance supply.[25]

In a forecast by the International Institute for Applied Systems Analysis (IIASA) and the World Energy Council (the common forum for the global energy industry), global investment in energy between 1990 and 2020 is projected to be around $12.4 trillion, or over $400 billion a year. If the average payback period for the loans involved is around 15 years, then total unpaid loans must amount to a constant $3 trillion or so. With such sums at stake, the large investment banks act as guardians of the grid-distributed energy industry and the pool of large-scale generating capacity. In Germany, the banks exert control not just through their lending, but also directly through seats on the supervisory boards of major electricity generation and supply companies.

The extremely close ties between the various links in the chain explain why the power station construction industry has always preferred multi-megawatt facilities, despite the fact that they could achieve the same or greater turnover with smaller power stations: they are following the demands of their customers, the generating companies. The strength of these supply chain relationships also explains why it is that the most ludicrous technological ventures are embarked upon with such great expectations – from the allegedly crucial fusion reactor to the crazy idea of setting up carbon sinks (in order to sequester excessive CO_2 emissions) while downplaying the potential of decentralized micropower plants. It also explains why electricity companies are prepared to make risky investments in waste disposal or telecoms, despite refusing to consider even the slightest risk vis-à-vis renewable energy, and also why certain 'suppressed inventions'[26] never even make it onto the market, for fear of rocking the boat.

Of the many individual steps taken on the road towards globalization and the formation of monopolies and cartels, most are economically sound and rational decisions, given the structures established by the fossil fuel and mineral industries. The situation as a whole, on the other hand, is becoming ever more irrational and problematic. Every incursion into the networks and supply chains of the fossil energy industry is portrayed as a danger, and dire warnings are given both to government and to the public. Change, supposedly, can only be the result of voluntary action within the cartel-dominated industry, which is why the response to climate change, for example, is not to regulate by statute, but at best to propose voluntary codes of conduct, which themselves are honoured more often in the breach. Of course, the energy companies publicly submit to the authority of the legislator, and draw their legitimization from the political and legal framework. But this only serves as a smokescreen for the embargo on all outside influence, whatever the consequences may be for the environment and the economic, social, democratic and international order. There are four developmental pressures at work in this process:

- the pressure towards economic globalization, which is the inevitable product of a fossil fuel-based energy system;
- mergers and acquisitions stemming from the need to integrate supply chains;
- cartel formation driven by industrial dependency on fractions of the same resource stream, focusing on the electricity industry, which is a consumer of all the fossil fuels, the supplier of energy in its most versatile form (electricity), the greatest single consumer of capital investment for its generating capacity and distribution network and the only industry to hold all other companies and citizens in its thrall; and
- the attempt by the electricity industry, as described below, to become a strategic economic lynchpin by moving into the provision of data and media networks, in which respect it has opportunities beyond the reach of all other industries.

Of the 50 largest European companies, 17 are in part or wholly conventional suppliers of energy and raw materials, or part of the chemicals industry. This does not include suppliers of power station technology, nor car manufacturers with their interest in cheap fossil fuels, nor the large food-processing companies with their close links to the chemicals industry and influence on primary agriculture, without whom an economy built on solar resources cannot be achieved. If you count all these, then 43 of the 50 are directly or indirectly involved in the established business of resource supply and processing, with an established interest in large quantities and low prices. The major banks do not figure in this list. So much for the 'weightless' economy.

The convergence of power: networking, supercartels and the disempowerment of democratic institutions

Since the advent of electricity, the business of its supply has always had an important role to play. Previously, the highest rollers have always been the oil companies, who have not brooked any public interference. In more recent times, though, the electricity industry has surged ahead.

The electricity supply is one of the essential infrastructure components needed to keep a modern society running, alongside the railway, the post office (including the telephone network) and the water supply. However, wherever regional electricity boards have been transformed into public limited companies, even where the state remains the majority shareholder, they have since begun using their regional monopolies as a springboard for moving into other companies and sectors where their activities are not so geographically restricted. This process has been underway for decades in Germany in particular, whereas the nationalized electricity industries of other countries have remained tied to their original business. The German federal monopolies commission has on many occasions criticized the expansion of the electricity industry into other sectors as a serious distortion of the marketplace.

Many believe this problem to have been solved with the liberalization of the electricity market. This is not the case in practice. The established electricity industry is now free to merge at will with the other transnational corporations within the resource supply chain. It also has a unique trump card in the potential multifunctionality of the electricity distribution grid. The grid is the electricity industry's passport for entry into the sector commonly thought to have the greatest importance for the future of industrial society, culture and democracy: telecommunications and electronic media in general. The stage was set with the dismantling of the state telecommunications monopoly and the existence of a thriving commercial television industry. The deregulation and privatization of the electricity, telecommunications and railway industries put an end to the classical division of labour among the formerly public-sector service corporations. The national grid could, in theory, also be used to transmit data; so too the overhead cabling of the railways. The retrofitting of telephone cables for high-voltage current, by contrast, presents considerably greater and more expensive challenges. In the struggle for control of these networks, the electricity industry has two blatant advantages that might allow it to steal a march on the former state telecoms and railways companies:

- Possession of the longest and most extensive network. In 1997, the primary grid in Germany had a total length of 492,000 km (307,500 miles), and 1,077,000 km (673,000 miles) of secondary cabling – enough to stretch 12 times and 27 times round the globe respectively. These lengths are considerably greater than those of the telephone network or the electrified rail network.
- Availability, through integration into the resource industries and the ability to tap their economic power, of the greatest reserves of capital and economic influence.

New applications for the existing network infrastructure, for which the EU Commission has coined the intentionally harmless-sounding term of technological 'convergence', are explicitly welcomed and promoted as a 'coherent concept' for all network and data-transmission services. The goal is greater asset productivity and combined service offerings: voice and data transmission over the internet, electronic business transactions and other online services, audiovisual feeds, the interaction of mobile and cable phone networks, the integration of computers and data storage into electronic services and the ongoing process of digitalization. As the development of IT makes it possible not just to use cable TV networks for data access and telephone services, but also to broadcast audiovisual content over the internet, the boundary between broadcast media and data transmission is dissolving. The capacity and bandwidth limitations of local landline networks still present a barrier to unlimited access to the internet and other online services. The case for convergent platforms for information and media services is argued not just in terms of economic efficiency, but also on grounds of customer and consumer convenience.

Clearly, such concentration of economic power brings with it a host of problems, including pricing difficulties, availability of content, network access and the difficulty of levelling the playing field. The common underlying issue is the question of vertical versus horizontal network integration – ie, whether a variety of network operators will compete as equals, or one operator will come to dominate. The policy preference, naturally, is for horizontal integration, but the signs are that vertically integrated operators will prevail, and the end result will be either oligopoly or monopoly. This seems especially likely since the electricity industry has made entry into the telecoms sector the centre of its long-term strategy, which it is realizing apace. The EU Commission report mentioned above, which summarizes the results of consultations with interested political institutions, companies and organizations on the need for regulatory action, contains the following passage:

> Many participants were of the opinion that competition regulations should be applied in the case of restrictive practices by existing network operators. There are fears of restrictive practices, illegal cross-subsidies and the bundling of access and content. Many network operators take the position that the application of competition law should take account of the high investment cost of rolling out digital communication networks and television platforms, and also of the great uncertainty as to the likely demand for these services.[27]

This last is an undisguised demand for political institutions to grant institutional and corporate investors a free hand. In the case of the power companies' attempt to break into telecommunication at least, this is exactly what is happening.

It is clear for all to see that multi-billion-dollar investments in this area are being cross-subsidized out of revenue from sales of electricity, in flagrant contravention of the regulations governing competition in the telecoms sector. Yet political institutions are not being half as rigorous in their application of new market regulations to the electricity industry as they are to the ex-monopoly telecoms companies. For instance, while the German telecommunications act right from the start contained provisions for a watchdog to monitor anticompetitive practices and guarantee unrestricted network access, the new German energy act has been drawn up on the basis of a voluntary sectoral agreement between industry and the electricity companies. The electricity companies, meanwhile, are busily gobbling up any remaining municipal and regional distribution grids not yet under their control. The ultimate aim is a convergent network under control of the electricity industry. In this battle of the networks, the likelihood is that the electricity magnates will win out.

If the politicians fail to apply the brakes on this development, then the future is clear. Privatization and the expansion of network power will be followed by the internationalization of the distribution grids. Power companies will become ever more removed from political control and market transparency, with the large-scale power stations being accorded a privileged position within the international electricity grid. Backed by the cartels of the fossil resource industry, the power companies will gain the upper hand over the convergent networks, a position from which they can exert control over electronic transactions and the media, and over television broadcasters in particular. With television broadcasters under their thumb, they will have the power to control public access to information and to shape public opinion. The outcome will be a supercartel of proportions unique in economic and political history, with the ability to withstand both market pressures and political institutions' futile attempts to regulate. Thus is the power companies' show of bowing to democratically elected politicians exposed for what it really is: a sham.

It will come as no surprise if in the near future we learn that media empires such as those owned by Murdoch,

Berlusconi or Kirch have been taken over by power companies. What can then hold back telecoms companies under the thumb of power companies from taking over IT services for all firms connected to the electricity industry by offering packet-switching services for both data and power? Who would then be able to check how often communication services contracts are tied to low network tariffs, preferential offers and targeted dumping? How can this be controlled on an international level when national controls fail, or were never even properly applied? This issue has not yet penetrated the public consciousness. Instead, the power companies are lauded for their activities.

Fossil-fuelled power corporations present more than just an acute environmental danger. Their control of electricity supplies and their influence on the mineral resources industry, plus the support of the large investment banks, makes them the most powerful element in the economy as a whole. They hold all the cards they need to construct a comprehensive commodity supply and media empire. They are closely bound up with the fossil fuel extraction and processing industry, and by extension with the chemicals industry. This latter has not only ensured that agriculture remains dependent on its supplies of fertilizers and pesticides. It is also harnessing biotechnology and patent law to massively deepen this dependency, and it has extensive links with the food-processing industry. The power corporations have links to the waste management industry, and are currently seeking to bring the municipal water utilities under their control. They are attempting to erect toll-gates on information and media networks. They are systematically taking over all the former public sector supply networks, but with no trace of public accountability or control. They are wreaking havoc on the environment, democracy and the free market.

Even if this is not their explicit intention, the power corporations are well on the way to becoming a uniquely powerful cartel. To this end they have no need of grand strategic visions. They merely have to follow, step for step, the economic logic of their existing supply chains. In this respect, their behaviour is as 'normal' as that of other firms; it is simply that the opportunities open to them and the resultant consequences are comprehensive and crushing. It is an unparalleled failure of political institutions that they not only do not oppose this development, but are even seeking to advance it.

2

Exploiting solar resources: the new political and economic freedom

ACCORDING TO ASTROPHYSICAL studies, the solar system, along with the Earth and the other familiar planets, will last for about another four and a half billion years. Throughout that inconceivable span of time, the sun will continue to give of its energies to people, plants and animals. Not only that, but the sun supplies energy in quantities sufficient to satisfy even the most opulent energy demands of an expanding human, plant and animal sphere. Every year, the sun delivers 15,000 times more energy than is consumed by the entire human population, to say nothing of the solar energy locked up in the short or medium term in the land surface, bodies of water and plant material. It is therefore utterly grotesque when people continue to assert – supposedly with scientific backing – that humankind's energy needs cannot be met from solar energy alone. It is still only a brave few who dare to suggest that renewable energy can supply all our energy needs, as the scenarios in *A Solar Manifesto* describe.[1] Many fear that if they address the possibility of completely dispensing with the conventional energy system, they will be ridiculed as scientific illiterates or naive idealists.

When it comes to solar power, even in this highly technological age, we have a decidedly pre-technological attitude. Although technology is normally thought capable of anything, it remains for most people inconceivable that it might achieve the relatively simple task of meeting energy needs from the sun. Technological hubris extends even to seeking to replace natural cycles and to influence highly complex evolutionary processes. Yet at the same time, despite all the known dangers of conventional energy, no interest is shown in investigating how the complete replacement of fossil and nuclear energy with other sources might be brought about, and this despite the fact that the technological feasibility of solar power has been demonstrated on numerous occasions. The argument that solar power devices produce less energy over their lifespan than is consumed in their manufacture never held water, and has been comprehensively refuted.

Everybody now knows of countless functional technologies for tapping renewable energy sources – even if only through media coverage – which have now long since been rolled out: photovoltaics (PV), windfarms, hydropower, wave power, tide power and biomass combustion for generating electricity; solar water heating and hot water storage tanks, heat pumps and biomass-fired boilers for heating; motors that run on liquid, liquified or gasified biomass; or hydrogen extracted using renewable energy for use as a fuel or to drive industrial processes. The World Bank has long since recognized the immense variety of technologies in its official publications; it has long since been possible to paint a comprehensive picture of a renewable energy future.[2] New and improved generation technologies will continue to be developed. It is already possible, on the basis of current technology, to work out to what extent, and with what geographical variations, solar generation technologies will have to be mobilized to meet humanity's energy needs.

Only three basic statistics are needed for this:

1 current aggregate energy demand;
2 capacities and footprints of individual solar generation technologies; and
3 insolation (in almost all cases, this can simply be looked up in an atlas), or prevailing winds, river flow-rates and agricultural and forestry land currently available or able to be brought (back) into cultivation, together with the spectrum and yields of available plants.

Feasibility is then merely an issue of the technical and organizational problems that the chosen combination of technologies present, in respect of the actual needs of a region or economy and the capacity and start-up costs of the various generation technologies.

In Germany, for example, the average insolation per square metre per year is 1100 kilowatt hours (kWh), and aggregate demand is around 500 billion kWh. Average annual output from PV (not to be confused with maximum efficiency under optimal conditions at optimal times) is currently 10 per cent of incident solar radiation (sunlight that reaches the PV panels) – roughly 100 kWh/m^2. It follows from this that to produce 500 billion kWh using PV alone, 5000 km^2 of solar panels would be needed. The sensible option would be to mount the panels on the roofs or walls of existing buildings. For Germany, this would mean that less than 10 per cent of construction surface – roofs, walls and motorway sound barriers – would need fitting with solar panels.

This calculation considers the ratio of plant footprint and quantity generated for PV alone – I am not seriously proposing that countries should meet their entire energy demand just from solar panels. A similar calculation can be performed for wind power. A 1.5 megawatt wind turbine – now the industrial standard – produces around 3 million kWh annually in areas of moderate wind speed. To produce 500 billion kWh, therefore, 166,666 of these wind turbines would have to be installed in areas of similar windiness. Of course, nobody with half a clue about renewable energy would even consider trying to meet all energy needs from wind power or PV alone. Constructive use of renewable energy requires a mixture of the various generation technologies – a combination not just of wind and PV, but also other sources, which nature offers in differing quantities in different locations.

The same simple method of scaling up the rate of introduction of proven technologies can be used to calculate the potential for renewable energy use in heating and cooling: how great is the energy demand in each case, and how many solar collectors or biomass-fired boilers would have to be installed to meet this need in a given climate?

Even in areas that receive less sunlight, such as central and northern Europe, there are already buildings which, in judicious combination with optimal insulation and heat exchangers, can be heated from solar radiation alone. There is therefore no rational reason why this could not potentially be applied to all buildings – and heating is where the greater part of energy consumption goes. In the case of motor fuels, the calculation of possible usage depends on which renewable energy source is or can be used in which region with what type of motor: vegetable oil, hydrogen produced using electricity from renewable sources, plant-derived alcohol, hydrogen or gas. The quantity of energy available from plant sources is a product of varying land fertility, the greatly differing energy content of various plants and their growth rates, of whether the whole plant or only a part of it can be used for power generation, and of the efficiency of the devices themselves.

Land currently in use for agricultural purposes totals around 10 million km². Around 40 million km² are covered with forest; the mostly unused arid and semi-arid areas add up to around 49 million km². Total annual photosynthetic output – ie, all plant growth, whether natural or for wood or food production – is currently around 220 billion tonnes of dry mass[3] (not to be confused with total mass). By contrast, 3.5 billion tonnes of crude oil, 2 billion tonnes crude oil equivalent (TOE) of natural gas and 2.4 billion TOE of coal – in total just under 8 billion TOE of fossil fuels – are extracted for use in electricity and heat production and as fuels and feedstocks for the chemicals industry. In the case of fuels, this represents almost the entire market demand, and for chemical feedstocks, by far the greater part of total demand.

Quick-growing trees can produce 15 tonnes of dry mass per hectare on average land with average water input; straw from cereals can provide between 12 and 18 tonnes, miscanthus over 30 tonnes, hemp 10–12 tonnes, and eucalyptus 35–40 tonnes per hectare.[4] Given an average yield of 15 tonnes of dry mass per hectare, less than 12 million km² of arable or forest land would be needed to replace the current consumption of crude oil, natural gas and coal for energy generation – assuming that the entire fossil fuel output were replaced solely by biomass grown exclusively for combustion, without adding arable crop residues or biogas evolved from organic waste. Biomass production, however, can be expanded – through afforestation, the cultivation of fast-growing, high-yield plants, and by using the whole plant as a source of energy and materials. There would be no need to encroach on land needed for food crops. Besides which, biomass cultivation need not compete with food crops for arable land, because less demanding plants can be cultivated across large swathes of arid land. The potential for afforestation is also great: it has been calculated that the 11 biggest tropical countries alone (out of 117) could expand their forest by 5.5 million km².[5] Again, as it would be nonsensical to try to meet all energy needs from biomass alone; here, too, there is a surfeit of solar resources waiting to be tapped.

All these calculations show that the limitations of restricted availability claimed for renewable resources do not obtain. In actual fact, the problems lie exclusively in the insufficient attention paid to renewable energy and the inadequate take-up of solar generation technologies. The calculations carried out here can in any case have only illustrative value, because the various sectors of energy consumption which have developed along with the fossil energy industry are all considered in isolation. A later section of this book will show that renewable energy sources allow the formation of completely different, integrated and far more efficient structures for energy use, which will render current distinctions between different energy sectors largely obsolete.

The solar supply chain

Energy needs can be met from renewable energy all over the globe, though of course the palette of available sources varies from region to region, country to country and continent to continent. The intensity of insolation, strength of prevailing winds, presence or absence of hydropower potential, forestry potential or availability and quality of land for biomass crops and level of precipitation will all affect the combination of sources which can be employed. It therefore follows that differing energy requirements in different regions will be met in different ways. It is this great structural variety which makes it difficult for energy ministers, who for decades have been used to the structures of fossil fuel supply, to get a feel for the potential of renewable energy. To recognize the economic, technical, cultural and political opportunities of renewable energy, it is insufficient simply to compare the outputs of individual generation technologies. Environmental impact can also not be assessed solely on the consequences of power generation itself – upstream activities are equally important. Isolated cost comparisons between fossil fuels and renewables conceal the breadth of the spectrum of applications for renewable energy. The key comparison must encompass the entirety of the respective supply chains.

A serious comparative study must also account for both constant and variable factors. The constant factor is the ultimate source of the energy. Renewable sources are not only far more diverse, but also widely distributed across the planet. The variable factors have to do with the differing technologies, which can always be improved, subject to the limitations of the relevant energy source, and with the economic cost of the energy produced. The universal availability of renewable resources within the local environment, however, opens up an opportunity which has been increasingly ignored since the Industrial Revolution, and which in consequence is today undreamed of: that of harnessing the source and generating the energy on the same site – or at least in the same region – where it will be used. The upshot of this is that the supply chains required to meet energy needs from renewable resources are much shorter – or even non-existent. With modern technology, this in turn holds out the possibility of regional or local energy self-sufficiency in place of the current global dependency on fossil fuels – an opportunity for new political, economic and cultural freedom.

The spectrum of opportunity is further enlarged by the possibility of replacing fossil fuel-derived materials with solar resources, which would allow regions with the right land and climate conditions to grow their own resource base. This would at least redistribute the resource base over a far greater number

Table 2.1 *Characteristics of solar resources*

	Theoretically inexhaustible?	*Risk-free source?*	*No geographical restrictions?*
PV electricity	Yes	Yes	Yes, with varying yield
Solar thermal electricity generation	Yes	Yes	No, depends on insolation
Wind power	Yes	Yes	No, depends on prevailing winds
Water power	Yes	Yes	No, depends on watercourse and the effects of deforestation and climate change
Wave power	Yes	Yes	No, depends on coastal situation
Geothermal power	No	Yes	No, depends on availability of underground heat
Ocean surface heat	Yes	Yes, albeit dependent on the effects of climate change	No, depends on geographical conditions
Solar heating	Yes	Yes	Yes, with variable efficiency
Air, ground and water heat	Yes	Yes	Yes
Biomass for energy generation and as raw material	Yes	Yes, if managed sustainably	No, depends on availability of suitable land

of countries, which might in turn lead to industrial relocation, changes in world trade flows and a new, more differentiated division of labour across the global economy. The overall potential and geographical availability of solar resources is listed in Table 2.1.

Obviously, incident sunlight and heat, wind and water power are not available equally in all parts of the world. Nevertheless, it is fair to say that what you've got, you've got. Only direct use of sunlight and heat irradiation is effectively unlimited, within the scope of local insolation. PV and solar heating thus offer the widest spectrum of use, and the technical availability is enhanced by the fact that the supply chain is extremely short. The other renewable energies such as wind and water power are geographically limited in scope; biomass is limited by the availability of arable land and appropriate plants, especially as provision for human and animal nutrition must be taken into account. In addition, biomass, be it food crops, energy crops or plant materials, can only be considered inexhaustible if the land used is not degraded by poor husbandry and woods are not simply chopped down without management or reforestation regimes. Even a resource that is in principle inexhaustible can be used up by extensive agricultural techniques.

Only the biomass supply chains have any appreciable length, as the energy source must first be planted and harvested. Even here, though, there are relatively few links in the chains in which production takes place directly at the point of end-use. The advantage of shorter supply chains is particularly relevant to electricity production from renewable sources. With the exception of biomass, the green electricity chain begins directly with electricity generation. These short chains bring a double environmental bonus: besides zero or only minimal environmental impact, the transport costs are far smaller. In economic terms, this means that capital costs for infrastructure are reduced, and that local economy can be revitalized.

Biomass

The first link in the chain is forestry or the cultivation of energy and resource-yielding crops. The second step is the harvest, when the raw plant material is also chipped or otherwise chopped up. Depending on its characteristics, the material is then prepared for combustion using specialized plant, for example, by pelletization or high-temperature gasification, or in some cases transported directly to a power station. This results in one or two further links in the chain. Ideally, biomass should only be transported short distances, because the lower energy content per tonne by comparison with a tonne of fossil fuel would otherwise render the transport costs too high. This is one reason why biomass power plants or other processing facilities should be located near the land where the biomass is grown. Another, more compelling reason is the necessity of returning nutrients taken from fields and woodland. It makes environmental and economic sense to spread the ashes from biomass combustion back on the fields or woodlands used to grow it, so that the nutrients they contain are not lost. Without thus closing the loop, biomass yields would be reduced, or the need to add artificial fertilizers would immediately weaken the bottom line. Nutrient return represents the fourth or fifth link in the chain. Gasified biomass is then transported to its ultimate destination, for use in power stations or to produce industrial process energy, which gives two more links. Gasified biomass may be shipped over longer distances than dry mass. Further links are required to transport the electricity generated, as described above. The supply chain for biomass is thus scarcely shorter than for fossil fuels.

If vegetable oil is used as an energy source, the supply chain comprises cultivation of oil seed, harvest, shipment to nearby oil presses, extraction of the oil and subsequent onward transport to power stations or industrial plants. Including nutrient return, the total chain length for vegetable oil is six links. The biogas chain is shorter: collection and delivery of organic waste to fermentation tanks, drawing off and transporting the evolved gas and subsequent combustion in power stations or industrial plant – four links in total. If this chain is extended through use of the fermented mass as fertilizer or pesticide, then additional costs result, but also environmental and economic advantages, the biogas chain being entirely local in scope. The number of links in the biomass supply varies according to use and processing technique, and the examples given do not represent an exhaustive list. There is quite considerable scope for replacing fossil energy and resources with plants, as detailed in Chapters 6 and 7, in the context of high-tech approaches to plant-derived energy and the resulting wide range of opportunities available.

With respect to this last in particular, it is scarcely possible to determine the number of links in a prototypical supply chain, as the materials and their uses differ too greatly. The chains may be geographically short but with many links – for example, where regionally produced wood is used as construction material – or cover long distances, as in the international wood trade, the production of lubricants from vegetable oil or the manufacture of plant-derived plastics. In essence, supply chains may be particularly short where the required plant material can be efficiently grown in the vicinity of the processing plant, but are of necessity longer where a highly specific material is required which grows only in certain regions, or can only be efficiently produced in the required quantities in those regions.

The basic principle, however, is the same: most countries must import fossil fuels and mineral resources for lack of their own reserves, whereas most solar resources can be produced domestically, as evidenced by countless examples of food crops initially native to one region, but now grown almost the world over. Potatoes and maize, for instance, were originally natives of the Americas; now they grow in almost every corner of the world. The same goes for rice and bananas, originally from Indochina, or beans from the Andes or wheat from central Asia. Of course, not all plant species can be so transplanted; local climates vary too widely. Nevertheless, the possibility exists for a relatively large number of species, especially where insolation, precipitation and land quality are similar. In any case, almost every region offers its own specific palette of useful crops.

Biomass remains unique among the renewable energy sources in being reliant on a supplier of primary energy. That being the case, it is entirely conceivable, despite the economic advantages to be obtained from proximity of production and processing to power plants, that biomass exploitation could follow the pattern of global business concentration and associated dependency relationships familiar from the fossil fuel industry. Indeed, multinational corporations are already buying up vast tracts of agricultural and forest land in order to secure their future position as suppliers of raw materials and energy. In this regard they are following the negative example of Brazil, where bioalcohol for millions of vehicles is produced from sugar beet grown in gigantic plantations. To get an idea of the potential risks, one only has to cast an eye over the food-processing industry, which has undergone a long-running and heavily internationalized process of business concentration, despite the requirement to produce locally enforced by the need for agricultural land.

The primary tools of would-be monopolists were direct production and supply contracts with agricultural enterprises. The first stage was to squeeze ever lower prices from the producers. Then came 'vertical integration', as the foodstuff companies moved towards direct control of agricultural production – dictating exactly which fruits the farmer should grow for optimal industrial use. This second stage followed on logically from the first, after the initially relatively independent farmers were forced into dependency or to relinquish their farms entirely. The third stage was the monopolization of plant and animal seed. Further expansion of this monopoly is sought through the patenting of genes, particularly in response to pressure from chemicals companies, a subject examined in more detail in Chapter 7. These developments were additionally facilitated by tax breaks for the international trade in agricultural produce and latterly by recent agreements on world trade, which do not distinguish between agricultural and industrial goods, as discussed in more detail in Chapter 9. Finally, the industrialized countries also grant direct subsidies to food exports. The WTO treaty does stipulate that these subsidies must be removed, the only positive aspect of WTO regulations in the agricultural sector.

Is it not then highly likely that these structures will also govern the increasing use of biomass as a source of energy and resources? It would not be the first time that the wrong approach has triumphed over economic and environmental sense because large corporations have used their overbearing influence on governments and parliaments to promote their own interests. Agriculture is one of the most prominent examples of this. If it were simply a question of securing the resource base for the post-fossil-fuel age, protecting the atmosphere from the trace gases emitted in fossil fuel extraction, processing and combustion, and overcoming the dependency of energy-importing countries on the very few exporters, then even biomass exploitation under the control of a small number of corporations would be preferable to fossil fuel use. In both cases the processing and marketing structures would be transnationally concentrated, but the advantages of biomass over fossil fuels would still be realized. However, a process of accelerated concentration would also bring dramatic social consequences for rural areas, run the familiar risk of reckless and extremely unbalanced farming methods and cause a global redistribution of nutrients equally fraught with environmental difficulty. The gigantic shipments of animal feed from the USA to Europe also export mineral nutrients, thus depleting US agricultural land and placing a strain on the capacity of European countries to absorb them.

The decisive difference between fossil fuel energy and fossil and mineral resources on the one hand, and plant-derived resources on the other, however, lies elsewhere. In the first case, the formation of global supply chains under the rule of transnational corporations is inevitable and irreversible; in the second, *global supply chains and transnational business concentration are by no means inevitable and, where this has occurred, reversible.* The determining factor is: who latches onto plant-derived resources and how? In the end, the politically determined framework for the energy and agricultural industries determines whether the exploitation of biomass for food, energy and raw materials results in long or short supply chains, industrial concentration or decentralization. In other words, concentration can be avoided, especially with radical reform of agricultural and energy policy. Where concentration and monopolization has taken place, this can be reversed as long as the land remains fertile or can be reclaimed by politically imposed land reform or regional market regulation.

Supply chains for industrial electricity generation from renewable resources

Where concentrated supplies of energy are available, it is more cost-effective to produce electricity using large turbines than using smaller power plants of the same type. Due to their wide distribution, this applies to renewable sources in only four cases:

1   biomass, which can fire power stations of up to 100 megawatts (MW) capacity;

2   highly concentrated quantities of water, such as strong river currents or straits, large natural waterfalls like Niagara, or artificial heads created by dams. The latter involves extreme interference in natural cycles, and is also problematic because of the risk of dam failure endangering entire regions;

3   tidal power, generating power from the ebb and flow of the tide in coastal regions; and

4   solar thermal plants, which either use collectors to concentrate solar heat to produce steam to drive turbines in the conventional way, or combine a large area of transparent film with a chimney-like tower to create an updraft through the tower. The updraft powers turbines mounted at the base of the tower.

In these cases, the electricity supply chain begins at the power station. As with fossil fuel or nuclear power, the electricity must then be fed through high-, medium- and low-voltage cables before it can be used to power lights and motors, giving a total of five links in the chain.

The supply chain for direct generation from renewable energy sources

The advantage of shorter supply chains is especially true of PV. Sunlight can be converted into electrical energy all over the world and with the lowest distribution costs of all generation technologies. As electricity is the most flexible of all secondary energies, suitable for the production of artificial light, powering machines and motors, heat pumps and refrigeration systems as well as for driving industrial processes, Harry Lehmann terms PV the 'prima donna' of the energy world.[6] It is only the (as yet) relatively high production cost of solar panels that distracts from their potential economic superiority.

In this case, the supply chain begins with the installed solar panel, which directly converts sunlight into electricity – without moving parts and thus almost without wear and tear, completely silently and without any emissions whatsoever. Of course, the panels must still be manufactured beforehand, a process consisting essentially of the production of the necessary material (currently predominantly silicon), cell manufacture, panel assembly, manufacture of the inverters to convert the direct current produced into alternating current, and lastly the installation of the panels themselves. But the production chain for the generating plant has not been subject to particular scrutiny in the analysis of fossil fuels either – neither for the power stations nor for the refineries or transportation. Where solar panels are used in isolation to produce electricity for consumption on-site, the generation process is then already almost compete, because the current need only be piped through the internal cabling of the house or device to power lights and machines. Electricity with no supply chain! There are only a few people who appreciate the fundamental importance of this.

Even where PV electricity is fed into the grid and thus embedded in a supply chain, the chain is still very short. The current need only be piped down low-voltage cables because it is generated in small quantities in numerous locations, which does not result in high voltages. Even power-station electricity reaches the end-user ultimately over low-voltage cables. PV electricity avoids the detours via the transformers and high-tension cabling necessary for fossil fuel energy.

The supply chain for wind power also starts at the wind turbine. If used autonomously, then, as with PV, there is only one link in the chain. Further links arise where the power is fed into the grid, depending on how often the current has to be transformed before it reaches its destination.

The supply chain for solar hydrogen

The use of electricity from renewable sources to electrolyse water into its constituent hydrogen and oxygen offers significant scope for expanding the spectrum of renewable energies. By this means energy can not only be stored, but the hydrogen can also be used as fuel to drive industrial processes or as a raw material for the chemicals industry. Discussions of these possibilities have hitherto centred on the heavy industrial approach – using large-scale solar power plants in the Sahara or large-scale hydro in Canada to manufacture the hydrogen, which would then need to be transported to Europe or the USA. Little or no consideration has hitherto been given to the separation of hydrogen in local small-scale electrolysis plants using renewable energy, or its extraction from biomass.

Yet this is precisely what must be considered, for otherwise the supply chain for hydrogen becomes as long as for fossil fuel or nuclear energy. The supply chain for centralized production begins with electricity generation in large-scale hydro or solar thermal plants. The electricity must then be fed over high-voltage cables to the electrolysis plant, where hydrogen is separated and subsequently liquified to render it transportable. The liquid hydrogen must then be stored in large storage tanks in the vicinity of harbours, before being shipped to its destination where it must once again be stored. Thence it can be distributed to power stations, garages and households, who place it in interim storage before it is finally used. That makes a total of 11 links in the chain, including the final conversion into energy. For regional production, by contrast, the chain is shorter, as electrolysis, liquifaction and storage can be co-located, and even electricity generation and electrolysis can be combined, which would render the development of a comprehensive hydrogen transport infrastructure unnecessary.

The economic logic of the solar energy supply chain

Fossil fuel and solar energy generation are intrinsically very different processes, and the opportunities they present for maximizing availability and efficiency – with respect to both resource consumption and financing strategies – are correspondingly diverse. Besides the differing environmental impact, the disparities between the supply chains demonstrate just how absurd it is to evaluate the economic potential of energy sources solely on the basis of the capital cost of the power generation plant required. It is because of such absurd reasoning that there has been such reluctance to exploit the potential of renewable resources.

Figure 2.1 compares the supply chains for fossil fuels and renewable energies, from which the following conclusions can be drawn:

- The shorter the supply chain – ie, the smaller the number of distinct processing steps involved – the greater the scope for reducing the costs of energy generation. If improved solar technologies can be introduced on a large scale, they represent not just the least environmentally damaging strategy for meeting energy needs, they are also potentially the most productive and thus the most economic solution. For this to happen, it is insufficient merely to recognize the benefits of solar energy. Technologies and strategies must be developed to exploit its advantages to the full. Insufficient progress on this front is the reason why the greatest potential economic benefit of renewable resources has not yet been systematically exploited.

    As long as they remain embedded within the conventional framework for energy generation, providers and consumers of energy from renewable resources will continue to pay the costs of fossil fuel supply and distribution networks. The potentially decisive advantage that renewable resources have over conventional fossil fuels will continue to go unexploited. If the switch to renewable resources simply replaces elements of the established fossil fuel structure, this will introduce a systemic bias that will hamper the growth of the renewables sector, confining it to a peripheral role within the energy industry for some time to come. Effective use of renewable resources requires a radical rethink of the supply and distribution network – simply copying the established structure will not work. The construction and operation of the distribution grid, for example, typically constitutes more than half the costs of an electricity supply. It is in the elimination of precisely these factors that the greatest opportunity for productivity gains from renewable energy resources lies.

    It follows from this that productivity gains from renewable resources cannot be realized through the construction of multi-megawatt power plants with sprawling distribu-

tion networks. That is not to say that there is no place for
solar thermal power plants. What is does imply is that such
plants should not be used as the core of an inter-regional
– or even international – distribution grid. The ideal use
for a solar thermal power station would be to serve large
towns and cities in its immediate vicinity – for example,
Cairo's power needs could be supplied by a plant located
in the nearby desert.

- On this basis, one criterion for evaluating the various
technologies available for exploiting solar energy will be
their potential for shortening or even completely eliminating the energy supply chain. On-site generation using PV
cells, for example, may potentially be far more economic
than large-scale generation plant.

- One decisive advantage for renewable energy in the future
lies in the ability to generate electricity at *minimal technological and infrastructural cost*. Because electricity is such a flexible
tool, the demand for electricity will grow at an increasing
rate, at the expense of other sources of energy.
Within the current system, it is simpler to supply fuel for
combustion when and where the energy is required.
Converting the same fuel into electricity requires additional
process steps, and thus is more laborious and technologically complex. With renewable resources, the opposite
applies: electricity generation using PV and wind turbines
is technologically the simpler route, whereas producing
combustible fuel is more complex and long-winded. This
reversal provides the template for the energy revolution to
come.

The complexity of fossil fuel and solar power generation

Renewable energy is regarded as uneconomic primarily because
of the allegedly greater material cost of local power generation
over centralized power stations. This reasoning is specious, as
it neglects to consider the long supply chains involved in fossil
fuel energy and the concomitant material cost of fossil fuel

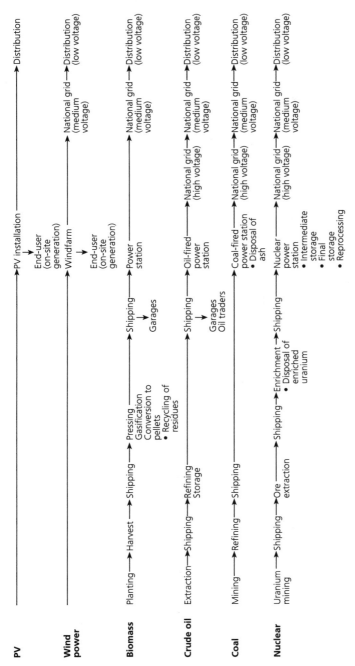

Figure 2.1 *Comparison of electricity generation from fossil fuels and renewables*

extraction and transport. It also takes no account of the relative complexity of the different generation technologies. Electricity generation from fossil fuels involves a considerably greater number of processing steps, resulting in proportionately greater technical costs (see Figure 2.2). In a fossil fuel power station, the first step is to convert chemical into heat energy through combustion (in a nuclear power plant the heat is derived from nuclear fission). Three further conversion processes follow: thermodynamic energy transfer to turn water into steam; conversion into mechanical energy as the steam drives the turbine; and finally conversion of mechanical energy into electrical in the generator. At the same time, the mechanical plant must also be cooled.

PV electricity generation, by contrast, involves only two steps: conversion of incident sunlight into direct current in the cell itself, followed by inversion to produce alternating current. In the case of wind power, the wind is converted into mechanical energy by the rotors, which in turn drive the generator to generate current. No cooling system is needed. Quite clearly wind turbine plant is not only easy to install; it is also more amenable to standardized production, and does not require operational personnel, aside from occasional maintenance work.

The fact that supply chains for solar power are short and the generation plant relatively simple really does beg the question of why generations of scientists and technicians have refused to accept it as an alternative, instead setting store by more laborious techniques, even preferring such extremely complex and technically fraught propositions as controlled nuclear fusion. Whereas complex technological solutions are placed on a pedestal, comparatively simple technologies are regarded with studied distrust, too backward for the modern, progressive age. Imaginative reservations are constructed in respect of 'simplistic' techniques, while justifications for high-tech approaches are grossly oversimplified.

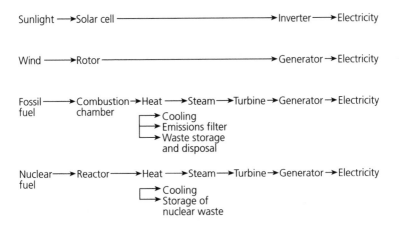

Figure 2.2 *Internal processing steps involved in solar and fossil fuel/nuclear electricity generation*

Short chains: the greater productivity potential of renewable energy

There are greater opportunities for productive energy use where power can be generated at a reduced technical cost. For renewable energy, however, these opportunities have hitherto been realized only in isolated niches – for instance, in low-end applications like pocket calculators, in the so-called 'passive use' of solar energy for heating and cooling buildings, in the solar collectors already installed on countless roofs in places like Greece and Israel, or in the 'solar home systems' in use in rural regions of developing countries, where generation from PV, even at this early stage in their development, is already more cost-effective than conventional power generation and distribution methods. Using PV saves on the cost of purchasing generators and the diesel to run them, not to mention erecting expensive overland distribution cables.

In the industrialized countries, such examples are regarded as side-issues or interim solutions for less developed countries. This dismissiveness means that the enormous opportunities that shorter supply chains present for increased productivity in the industrialized countries are all too quickly overlooked. Of course, for these opportunities to be realized, there must first be a conceptual break with the idea that what worked for fossil fuels will work for renewable energy. The structures required for fossil fuel energy are not compatible with the economic requirements of an electricity supply based on solar energy.

Solar power: technology without technocracy

With very few exceptions, centralized power generation from solar sources would effectively negate the advantages accruing from a decentralized energy supply (the subject of Parts III and IV of this book). It would be daft to close down existing hydropower dams; large-scale solar thermal plants to supply tropical cities make sense. Equally, the resulting disconnection from the regional resource base makes biomass-fired power stations of over 100 MW nonsensical; nor are large tidal power plants along great stretches of coastline to be recommended, when wind power provides a simpler, less costly solution more in keeping with the natural landscape.

In any case, even centralized solar power generation in sunnier parts of the world would inevitably need to be supplemented by local PV, wind, small-scale hydro and biomass plant. Yet the history of power generation and supply teaches us that although large-scale generation plant may be technically compatible with small-scale plant, combining the two produces structural conflict. Operators of large-scale plant need to run at full capacity to recoup their costs; the unpredictable output of small-scale producers is an irritant. There is little reason to assume that operators of large-scale solar power plant would behave any differently towards smaller suppliers than operators of nuclear and coal-fired plants. In the case of fossil fuels, the obsession with large-scale production and supply is a reflection of the underlying economic realities. In the case of solar power, the same obsession would be ill-considered, an expression of the industrial fantasies typical of the 20th century.

Grand schemes following the pattern of concentrated generation and supply set by fossil fuel power generation have been and continue to be entertained in respect of renewable energy. They are the product of a paradigm that can conceive only of individual large-scale remedies for large-scale problems.

One example of this was the proposal of the Munich-based architect and author Herman Sörgel, first presented at a 1931 exhibition of architecture in Berlin, to construct a dam right across the Straits of Gibraltar to control the flow of water from the Atlantic into the more low-lying Mediterranean. The idea was to build a gigantic hydropower plant to supply Europe with electricity, while at the same time lowering the level of the Mediterranean to reclaim additional land from the sea along the Mediterranean coast and create a land bridge between Europe and North Africa. The Adriatic was to become dry land; Naples would have ceased to be a port. The project was much talked about; it fascinated Hitler as much as Mussolini.[7] No thought was given to the incalculable consequences of reshaping the Mediterranean ecosystem on such a scale; the project was too tempting to geopolitical ambitions of making North Africa both a part of and, in conjunction with the greening of the Sahara, the breadbasket of Europe.

Another, more contemporary example is the GENESIS project (Global Energy Network Equipped with Solar Cells and International Superconductor Grids) that some minds are toying with. The idea is to construct a global belt of linked solar power stations running along the equator to supply the entire world energy demand through a superconducting distribution grid. The supposed advantage would be an uninterrupted supply of solar electricity, because the difference between day and night and seasonal variation in output between hemispheres would cancel each other out.[8] But the result would be a hypercentralized global energy supply, the global dominance of one generation technology with the longest supply chain imaginable and colossal infrastructural costs. It is a product of technological megalomania with absolutely no conception of the sociopolitics of energy supply.

A third and similar example is NASA engineer Peter E Glaser's concept of an orbiting 'solar farm', which also crops up in discussion from time to time. Electricity for all Earth's inhabitants would be produced from PV platforms, orbiting the Earth, with a total surface area of many square kilometres, free of the limitations of diurnal and seasonal cycles. Generative efficiency would be very high because even incident sunlight outside the angle of refraction (a product of the Earth's curvature) could be exploited. The current produced would be beamed back to the ground in the form of microwaves with a ground footprint several kilometres in diameter. These would be picked up by a ground station almost 200 km² in size, converted back into electricity and fed into the distribution grid.[9] The same applies to this concept as to the GENESIS project: it may be technically feasible, but otherwise impractical, with no consideration of risk, economic viability or social consequences, and a failure to appreciate the real opportunities that solar power presents.

Proposals that turn a local resource free of supply chain dependency into a hypercentralized generation and supply system, maximizing dependency, are the product of a technocratic approach that has no regard for social context; an approach which, even without such engineering mega-projects, has already led to the shaping of society to fit technology, rather than – finally – adapting technology to meet real needs. Even the idea of using Saharan solar power to produce hydrogen for export, although on a considerably less monstrous scale than GENESIS project or solar power satellites, fails to do justice to the economic, social and political dimensions of solar power. A resource that is universally available across the planet without recourse to extended supply chains need not and should not for any reason be first centralized under oligopolistic or monopolistic business structures before being redistributed to consumers at large. We must learn and understand that it is not necessary to take circuitous and technically complex routes when there are direct, simple solutions available. It is only possible to understand and harness the social and economic capacities of technology if we take an untechnocratic approach to it.

De-monopolization and re-regionalization through solar resources

The dynamics of the drive towards business concentration is thought to be the dominant force in economic development, and, indeed, more and more industrial sectors are going down this route, even those for which concentration is not imposed by the resource base, as it is in the case of the fossil fuel and mineral resource industry described in Chapter 1. Many economic analysts therefore assume that the introduction of solar resources will also be followed by a process of business concentration, and in consequence many also view local installations of solar generation plant as merely the precursor to a development whose end-point will be solar resource plantations in areas of high insolation such as North Africa. In actual fact, the scope for concentration with a solar resource base is limited. Indeed, the dominant force may well be the very difficulty of monopolizing solar resources, thus turning conventional and seemingly universally applicable experiences of economic processes on their head. Mathematically speaking, it follows that the equivalent to a 1000 MW power station would be – depending on their individual ratings – 2000–4000 wind turbines, 1 million solar panels, or 50 large or 5000 small biomass plants; in practical terms, the equivalent energy production would be achieved using a combination of these sources.

The difference between the conventional energy industry with its four corporate pillars – the oil, coal, gas and uranium extraction and trading companies, the power station operators and the (in most cases identical) operators of the distribution grids, the power station construction industry and the investment banks that underwrite all the above – and renewable energy is that, in the latter case, only one sector is exposed to concentration and monopoly: the manufacture and construction of plant (ie, solar collectors, solar cells, wind turbines and biomass plants).

If renewable energy sources ever come to dominate the market, then the rump of the industrial webs described above, the fossil fuel extraction and trading companies, will slowly dwindle away. There will be nothing to replace the niche currently occupied by companies that extract or supply fossil fuels if fossil fuels come to be displaced by solar heating, sunlight, wind, waves and water currents. As Franz Alt very neatly puts it, 'the sun sends no bills'.[10] The basic problem that fossil fuel companies have is that sunlight and wind cannot be patented and sold under licence. Comprehensive use of renewable energy would take the wind from the sails of an economic globalization and industrial concentration process driven by the scarcity of fossil fuel reserves. This alone would spark a process of de-concentration, de-monopolization and the re-regionalization of economic structures.

The two spiders in the fossil energy industry web – the operators of power stations and electricity and gas distribution grids – will also have no further role to play in a decentralized energy supply based on solar power. Large power stations need large companies to run them; small local plants have no such need. Once the transition to electricity supply from renewable sources can no longer be stopped, the power companies will naturally seek to gain control of these sources. In the case of PV, the highly decentralized nature of the plant makes this an essentially futile exercise. They will have more success with wind, especially with windfarms and offshore installations in coastal waters – how much success they would have depends on the extent to which the laws regulating the energy market favour this. But as generation plant for renewable energy is subject to natural limitations – the effective maximum capacity for individual wind turbines, for example, cannot be much more than 5 MW – power plant operation will no longer be purely the preserve of large companies. Provided that the market is freely accessible, many new types of enterprise are likely: local enterprise, on-site generation by companies, producer cooperatives and innumerable individual suppliers on a regional and local level. The politically and economically explosive potential of renewable energy is its universal availability, as this eliminates the dependency of both society and political institutions on power companies and reduces the influence that those companies can exert. Every large-scale power station decommissioned, every new local plant constructed and above all every improvement in power storage technology reduces the central role played by the national grid, to the point at which it becomes superfluous.

The economies of scale, which have favoured concentrated business structures because of their ability to mass-produce cheap consumer goods and so squeeze out smaller producers, do not apply to renewable energy. The rise of renewable energy disrupts two of the fossil fuel spiders' strongest webs, and the third web, the dominant role played by the large investment banks in the energy industry, is at least weakened. In a decentralized market, all potential investors, not just banks, can be sources of finance; the large investment banks will be just one player among many.

As mentioned, the market for renewable energy plant remains open to concentration and monopolization. It is possible that, following an initial boom, the global market for solar panels and accessories, solar collectors, wind turbines and biomass plants will come to be dominated by a very few firms. For the power station construction industry, this could even present a golden opportunity for diversification, provided they can make the leap from catering to a few large clients to serving many small ones. For PV and solar collectors, the customer base will be larger even than that of the car industry. That notwithstanding, manufacturers of solar generation plant will not be able to completely dominate the market. They will be dependent on a multi-billion-customer client base with a diverse demand structure for various panels and integrated systems. There will be scope for a broad spectrum of manufacturing and distribution firms, and an even broader palette of technical engineering and installation services.

The representatives of the fossil energy industry have been written out of the script for the renewable energy story, or allotted at most a secondary role; the market for renewable energy will no longer have a niche for conventional sources – at least, not with turnover at high as it is at present. Conventional energy companies are bound to old fossil fuel structures by the sheer scale of their investments; their business models, based on large-scale industrial plant, will prove their own undoing in the transition to renewable energy. A solar resource base makes it impossible to retain or ever re-create the power structure that has hitherto prevailed in the energy sector. The extent to which industrial concentration and monopolization is inevitable with fossil fuels and avoidable or impossible with solar energy is compared in Table 2.2.

Table 2.2 Can industrial concentration and monopoly structures be avoided?

	Nuclear power	Coal/gas/oil	Biomass	PV	Wind power	Small-scale hydro	Solar heating	Large-scale hydro/solar thermal plants	Hydrogen from solar power (heavy industry)	Hydrogen from solar power (light industry)
Primary energy	No	No	Yes	–	–	–	–	–	–	–
Trade in primary energy	No	No	Yes	–	–	–	–	–	–	–
Processing	No	No	Yes	–	–	–	–	–	No	Yes
Plant manufacture	No	No	No	No	No	No	Yes	No	No	No
Power station operation	No	No	Yes	Yes	Yes	Yes	–	No	Yes	Yes
Power distribution/trading	No	No	Yes	Yes	Yes	Yes	–	No	No	Yes
Plant finance	No	No	Yes	Yes	Yes	Yes	Yes	No	No	Yes

No: Industrial concentration and monopolization are inevitable (except for combined heat and power plant).
Yes: Industrial concentration and monopolization are technically and politically avoidable or impossible.

The short supply chains for renewable energy sources will end the pressure to globalize that comes from the fossil resource base. The dense interconnections between individual energy companies and between energy companies and other industries that result from fossil fuel supply chains will no longer be necessary. Shorter renewable energy supply chains also make it impossible to dominate entire economies. Renewable energy will liberate society from fossil fuel dependency and from the webs spun by the spiders of the fossil economy.

PART II

THE PATHOLOGICAL POLITICS OF FOSSIL RESOURCES

If the world continues to see fossil resources as indispensable and the alternatives as unrealistic, then it will continue to shrug off the dire consequences as unavoidable. The matter barely seems to merit a second thought. While the environmental consequences of energy use, resource costs, questions of efficiency and productivity and the longer-term availability of energy sources are all topics for discussion among politicians with environmental and energy portfolios, the standard response is to seek fossil fuel replacements for fossil fuel resources. The current focus is on natural gas as a replacement for crude oil, coal and nuclear power. Renewable energy is seen as a minor player in energy provision, and renewable resources are thought even less important.

That goes not just for the industrialized world, but also for the industrializing economies of the developing world. The industrialized countries' migration to ever more complex industrial technology, larger-scale supply systems and different patterns of energy and resource use has taken place gradually, over many decades. In the developing world the transition to modern energy systems has come as a sudden jump from a rural agrarian society to centralized industry. Developing countries have consequently suffered much more acute social and economic stress than the established developed countries. Nevertheless, this stress is generally not perceived as intrinsic to a fossil fuel and mineral resource base. The myths of the fossil fuel and minerals industries bamboozle political and economic leaders into overlooking the obvious.

CHAPTER **3**

The 21st century writing on the wall: the political cost of fuel and resource conflict

THE COSTS OF fossil resources are generally evaluated solely in terms of their market prices. Yet if we are finally to take on board the other, far more problematic side of the equation, then we must include the many and various incidences of environmental damage arising from the consumption of coal, gas and oil. Environmental costs being difficult to calculate exactly, almost every concrete figure has been successfully shot down. One thing, however, is clear: the environmental costs are so vast that it would be irresponsible to ignore them simply because it is difficult to pin them down to a precise figure. On top of the environmental costs, there are also political costs, which although even more difficult to quantify can be described and estimated in political terms.

The visible ravages of the fossil fuel economy are the environmental writing on the wall, bearing ill tidings for the 21st century. By overexploiting the planet's resources, we are sowing the seeds of a bitter harvest. It is a harvest that will be reaped at different times in different places, and which therefore threatens to bury the One World ideal forever. It is a harvest that will reshape the values of every society it touches. The conflicts of the 21st century could well escalate into a 'clash of civilizations', as US political scientist Samuel Huntington predicts.[1] Yet Huntington fails to see the root cause of these conflicts: the deep-seated need to secure a potentially endangered fossil resource base.

Many have expressed the hope that humankind's common dependency on fossil resources would be a force for peace, by compelling the economic powers of the world to cooperate. This is naive and wishful thinking. Resource reserves are in truth the flashpoints for ever more (and, in all likelihood, ever more acute) conflicts. Crisis and war have thus far remained confined to isolated geographical regions. Yet Iraq, Chechnya and the like are but a foretaste of a gathering conflict that threatens the very existence of global civilization. The following sections assess the threat posed to world civilization by the impending exhaustion of its fossil resource base.

A world in denial: the disregard for limited reserves

The gulf between political promises and the situation on the ground is as breathtaking as it is true. Despite all warnings of environmental damage resulting from irresponsible energy use; despite speeches promising action and national and international resolutions to curb energy consumption; despite all the progress towards less energy-intensive technologies; and despite dwindling reserves, global energy consumption continues to rise, and indeed to rise faster than ever before. With seeming inevitability, we are nearing a critical juncture where demand for fossil resources can no longer be met, yet where suppliers nevertheless continually strive to expand their markets. According to statistics prepared by the International Energy Agency (IEA) of the Organisation for Economic Co-operation and Development (OECD), between 1971 and 1990 the commercial energy supply rose from 4.9 billion TOE to 7.8 billion – an increase of around 60 per cent. Predictions for the period 1990–2010, based on a medium-growth scenario, suggest a further increase to 11.5 billion TOE (a rise of 48 per cent), growing to 13.75 billion in 2020 (an increase of 77 per cent on 1990).[2] Let us remind ourselves: the target set at the Earth Summit in Rio de Janeiro in 1992 was to limit climate-damaging energy consumption to 1990 levels!

One might criticize these predictions for not taking account of possible efficiency gains. But against that must be set the fact that the IEA forecasts expected price levels to be higher than they actually turned out to be. For the period 1998 to 2010 an oil price of $17 a barrel was assumed; the actual price in 1999 was under $13 a barrel. And as electricity prices are sinking in the wake of the liberalization of the electricity markets, it is highly likely that energy consumption could grow even faster than the IEA expects. The only effective spur for fast, clear improvements in energy efficiency is increasing prices. Yet against the backdrop of global competition, price increases are decried as a nightmare scenario and have become politically taboo. Global energy markets come before the fate of the globe. In view of the environmental consequences, such twisted priorities turn market economics into extremist dogma.

The forecasts show the lion's share of the increase in energy consumption falling to fossil fuels, with atomic energy taking a slightly increased share of the cake and only minimal growth for renewable energy. The IEA is a joint institution of the OECD governments, and its predictions serve as guidance and justification for fossil energy investors. Although the OECD governments have pledged themselves to reducing their energy consumption, there has been no protest at the exorbitant increases predicted by their joint energy agency. The OECD governments thus stand revealed as disingenuous charlatans. The same goes for the EU, so quick to denounce the USA as the greatest squanderer of energy in the industrialized world and also keen to point out the large growth in energy consumption in the developing world. Yet energy consumption is far from falling among the European OECD countries, who are in the main also the EU member states. Total energy consumption was 1.15 billion TOE in 1971, had risen to 1.43 billion TOE by 1990, and is set to rise to 1.95 and 2.05 billion TOE in 2010 and 2020 respectively. Between 1990 and 2010 alone, this represents an increase of 36 per cent, rising to 43 per cent by 2020.

Even on the global scale, the IEA forecast assesses the contribution from renewable energy as marginal. Production from renewable sources was 110 million TOE in 1971, had risen to 218 million TOE by 1990, and is only expected to reach 379 and 465 million TOE by 2010 and 2020. If these predictions actually came to pass, then the gap between commercial energy supply from conventional sources and from renewable sources would be drastically widened, rather than reduced. Quite clearly, despite the acknowledged huge dangers, we are happily steaming ahead on our previous course. This can also be seen in the figures for global CO_2 emissions, which were 21.4 billion tonnes in 1990, and are expected to rise to 31.2 billion tonnes by 2010 alone. That would be an increase of 46 per cent in place of stabilization at the – already dangerously high – levels of 1990.

It is by no means just the economic growth spurts of China and India which are to blame for this. Equally guilty are the rapidly growing global transport industry and rising electricity demand – the two sectors which are particularly closely bound up with the political priorities of economic globalization and modernization. In 1971, fuel consumption accounted for only 22 per cent of global energy demand. In 1995, it was 26 per cent, and it is expected to reach 28 per cent by 2010. Electricity made up 22.7 per cent of global energy consumption in 1971, rising to 26.8 per cent by 1995, with 29.8 per cent predicted for 2010 – not least as a consequence of the growth in power-hungry information technology. As electricity is overwhelmingly generated in large steam-turbine power stations, and as these have priority in investment plans (despite the fact that they produce the greatest losses in the conversion of primary to secondary energy), every increase in fossil fuel electricity use flies in the face of climate change obligations. Despite recent and anticipated improvements in the efficiency of generation technologies, the result has been a fall in aggregate efficiency. In 1971, end-user energy consumption accounted for 74 per cent of aggregate primary energy input. By 1995 this had fallen to 69.5 per cent. At the same time, the industrialized countries are becoming increasingly dependent on imported primary energy.[3]

Even if no environmental damage were to result from fossil fuel use, or if its effects had been exaggerated, the increase in fossil fuel consumption nevertheless has alarming consequences, as it dramatically accelerates the rate at which fossil fuel reserves are exhausted. Yet the faster we approach the point of resource exhaustion, the more strenuous the denials that this economic blind alley exists – despite the fact that the vital statistics on fossil fuel reserves have been available at least since the Global 2000 report prepared for President Carter in 1980.[4] There has been little to change these statistics since then, aside from the fact that the estimated rate of growth in energy consumption was considerably in excess of what actually happened. But this one change was sufficient for the fossil fuel industry, and the governments who accept its analyses apparently without question, to sound the all-clear. The public is losing patience with repeated and seemingly exaggerated warnings. Psychological reasons alone are cause enough to grasp at every straw to offer the comforting illusion that it won't be as bad as all that after all. Any good news – for example, of newly discovered reserves – is gratefully received, even when the amount found makes only a marginal contribution to world energy consumption. Thus when in December 1997 the papers reported that the French oil group Elf had found a 'giant oil field in Angola' containing 730 million barrels, they forgot to mention that this is equivalent to only ten days' oil supply! Another reason for denying the facts is that they simply do not fit either business or macroeconomic models of energy-centric economic development; in particular, they upset plans to unleash the economic potential of the newly opened global market. Discussions on the impending scarcity of fossil fuels are an irritant to the business psychology of the global economy, whose movers and shakers are betting on the accelerated industrialization of the developing world and who have their eyes on Russia's resource reserves. Everything points to the greatest growth conflict in world history. This conflict will also most likely be the last, and its aftermath will be chaos.

The limited reserves of fossil energy

As statistics on remaining reserves of fossil resources tend to vary, most people see little reason to believe the most far-reaching estimates. Yet commonly cited measures of how long various resources can last could equally well prove less than experts estimate. Countries with extraction industries tend, for example, to overstate their reserves in order to obtain proportionally higher extraction quotas at the annual OPEC quota allocations; greater reserves also improve international creditworthiness. Nevertheless, estimates of crude oil reserves are no longer as divergent as they were (especially as it does not particularly matter whether supplies will last perhaps only another ten or twenty years longer), and there is no room for great hopes. As the writers Jörg Schindler and Werner Zittel laconically observe: 'Pretty much everything has already been found.'[5] According to a survey by the German Federal Institute for Geoscience and Minerals (BGR), estimates for confirmed oil reserves range from 118 to 151 billion tonnes:[6]

- United States Geological Survey: 118 billion tonnes;
- World Oil, Annual International Outlook: 132 billion tonnes;
- Oil and Gas Journal: 138 billion tonnes;
- BP Statistical Review: 141 billion tonnes; and
- BGR: 151 billion tonnes.

Another all-clear went through the media in July 1999, based on the BP-Amoco Statistical Review of World Energy 1999. According to this report, in 1998 world energy consumption had sunk for the first time in 16 years, whereas world oil reserves were up 1.5 per cent on the preceding year, to 141 billion tonnes. A closer examination of the report revealed, however, that in actual fact, as a result of the crisis in Asia and low economic growth in Europe and the USA compared to 1997, growth up to that point had declined by merely 0.1 per cent. And the 1.5 per cent of additional oil reserves were equivalent to no more than seven months' supply, well within the range of the above estimates.

Estimates that go beyond the above also include so-called non-conventional oil reserves: heavy oils, tar sand, oil shale or oil fields in deep waters or in the polar regions. The potential for such sources is overestimated, as the extraction costs are extremely high, rates of extraction low and the environmental cost horrendous. Whereas oil fields require only a single well, in the case of tar sand and oil shale the entire mass of earth or shale must be dug up and washed and the oil pressed out. The process is comparable with lignite mining, and the energy loss due to extraction is similarly high. Colin J Campbell of the Geneva-based Petroconsultants, one of the most respected consultancy firms in this area, concludes that total practically accessible reserves total around 180 billion tonnes, and extraction rates will be in constant decline from 2000 on.[7] Even if annual oil extraction remains at the 1995 level of 3.32 billion tonnes, this means that oil reserves would be exhausted around 2050. With the US Geological Survey estimate of 118 billion tonnes, this point would be reached as soon as 2030. If the IEA forecast is correct in predicting annual oil consumption of 4.46 billion tonnes by 2010, then even by Campbell's calculations, oil reserves will be exhausted by 2040. The IEA, however, is assuming annual oil consumption will have risen to 5.26 billion tonnes by 2020 – which, in view of the strong growth in car manufacture and the exorbitant increase in freight traffic, and in air freight in particular, is by no means unrealistic. This means that the oil could be gone as soon as 2035, even under optimistic estimates of remaining reserves. The political alarm bells will in any case be ringing much sooner than that, because the crisis point is reached well before the last drop is consumed.

Estimates for remaining natural gas reserves, again according to a survey by the BGR, vary as follows:

- United States Geological Survey: 131.8 trillion m^3 (cubic metres);
- World Oil, Annual International Outlook: 144 trillion m^3;
- Oil and Gas Journal: 144.3 trillion m^3;
- BP Statistical Review: 144.7 trillion m^3; and
- BGR: 152.9 trillion m^3.

If annual extraction rates remain constant at 2.3 trillion m^3, this equates to around 57 to 65 years' reserves. It is, however, precisely in natural gas consumption that the greatest growth rates are seen, which means that reserves could be exhausted as soon as 2040. There has been speculation concerning non-conventional reserves,[8] but there are also many reasons for not including these in estimates of remaining reserves. There are estimated to be 2 trillion m^3 locked up in permeable rocks, but this is the equivalent of just one year's supply, scarcely enough to pay the immense cost of the new extraction techniques that would be required. Coal seams are thought to contain around 130 trillion m^3 of natural gas, but tapping these reserves is contingent on complete extraction of all coal reserves, which would massively increase the already unconscionable environmental risks. The extraction of up to 10,000 trillion m^3 from sedimentary basins deep in the Earth's crust has yet to be seriously costed. Speculation also focuses not least on natural gas from hydrates, compounds of gas and water locked up in the permafrost regions of Alaska, Greenland, Canada, Russia, the Antarctic and along the continental shelf, estimated at 1000 trillion m3. In order to gain access to these reserves, however, treaties such as that on the Antarctic would have to be torn up, or enormous capital investment would have to be made, to say nothing of the grave risks associated with such major disturbance of oceanic ecosystems.

Coal reserves are estimated to be around 560 billion tonnes,[9] enough to last 169 years at the current rate of extraction. However, coal consumption in particular cannot possibly remain constant as long as there is no decisive shift towards renewable energy. Annual consumption in 1995 was 2.35 billion tonnes; according to the IEA, this is set to increase to 3.3 billion by 2010 alone, and to 3.95 billion by 2020. This would reduce the lifetime of coal to 123 years. Then there is the massive increase in coal consumption that will result from the exhaustion of oil and gas reserves. If the energy industry continues to rely on fossil fuels, at that point coal would have to satisfy not just the demand of power stations and heating systems, but meet the need for motor fuel, with the aid of coal gasification and liquifaction. On this basis, coal reserves would be exhausted considerably before 2100. It is true that on top of the reserves already mentioned, there remain an additional estimated 500 billion tonnes of brown coal, but as Chapter 1 explained, lignite can only be used if it is processed and consumed directly at the point of extraction. For this reason, lignite reserves are not usually included in international statistics. On top of which, brown coal produces by far the greatest emissions of all the fossil fuels. The complete consumption of all remaining brown coal reserves would place a massive additional strain on the climate.

Annual extraction of uranium is currently around 60,000 tonnes, which even according to figures produced by the World Energy Council in 1993 gives a lifetime of only 41 years – into the mid-2030s.[10] Uranium reserves are subdivided according to the different extraction techniques and costs involved. Reserves costing up to $80 a kilo are estimated at 2 million tonnes. A further 2 million tonnes fall into the cost band of up to $130 a kilo, and for costs above that there are speculative figures of between 8 and 10 million tonnes.[11] It is, of course, questionable whether there will be any further increase in consumption, as more and more countries are moving away from nuclear power. No new reactor has been built in the USA since 1973; there has not been a new reactor in Germany since 1987, and the industry is due to be wound down. Sweden, Switzerland and Canada are also winding down their nuclear industries. How many of the replacement and new reactors in Russia will actually be built and commissioned is equally as uncertain as reactor plans in Japan, China, India, Brazil and several other countries. France, the UK and Belgium have no plans for new reactors. The cancellation of the Kalkar fast breeder project in Germany, the shut-down of the Superphénix reactor in France after less than 200 days' operation and bad experiences in Japan have scotched plans to continue and extend the lifetime of atomic energy through the introduction of fast breeder reactors. Other dreams for a nuclear age are also beginning to fade, such as the transition from nuclear fission to nuclear fusion promised for 2050. As governments are becoming less ready to stump up the double-digit billions required for the next reactor prototype, the fusion project is withering on the vine.[12] The nuclear industry, though, is nevertheless betting on the exhaustion of fossil fuel reserves and that renewable energy will not take off – in which case the phoenix could indeed rise from the ashes.

The limited reserves of mineral ores

The term 'non-renewable resources' includes both fossil hydro-carbons used in the chemicals industry and mineral ores. Reserves of the former are the same as for crude oil, natural gas and coal. Around one third of fossil hydrocarbon output is used as chemical feedstocks as opposed to fuel; only through drastic reductions in fossil energy consumption could the lifetime of hydrocarbon feedstocks be lengthened. Mineral ores, by virtue of their variety, partial mutual interchangeability and opportunities for recycling, offer much greater flexibility than fossil fuels. Mineral ores are important to the production of highly specialized materials, such as those used for high-temperature applications, aviation or military technology.

Despite all this flexibility, supplies of certain metals will be limited in the foreseeable future, as Table 3.1 shows. Recycling is not always possible – it depends on how the ores are processed and what metal compounds result. For most metals, there is always the option of exploiting even low-density deposits, but these are much more costly and energy-intensive to extract and process.

Table 3.1 *Mineral reserves*

Mineral	Reserves (definite and probable), in 1000 tonnes	Global output in 1996	Estimated duration, in years
Bauxite	22,983,000	114,000	202
Lead	63,400	2912	22
Copper	311,500	11,006	28
Nickel	35,814	1051	34
Zinc	143,200	7283	20
Tin	7190	196	37
Iron	68,880,000	549,000	125
Chrome	1,496,000	12,000	127
Manganese ore	875,600	8000	113
Cobalt	11,615	23	499
Molybdenum	5645	128	44
Niobium	4513	16	182
Tantalum	25	0.4	65
Vanadium	7480	35	213
Tungsten	2244	32	70
Gold	37	2.3	17
Silver	288	15	19
Platinum metals	57	0.3	198

Source: BGR[13]

Dwindling reserves versus worldwide growth in demand

If productivity gains lead to reductions in demand, then the foregoing figures will need to be revised. Nevertheless, in the case of fossil energy consumption, it would be crazy to move from draining the last drop of oil to squeezing out the last lump of coal from the bowels of the Earth. Not just in the light of scientific research on climate anomalies and the risk of catastrophic climate change, but also in consideration of evolutionary history. It was only with the arrival of photosynthesis, which allowed carbon to be sequestered in the form of plants, that the oxygen vital to the development of aerobic species was released. If this oxygen is reabsorbed through combustive oxidation of the sequestered carbon, then aerobic lifeforms will be deprived of the air they need to survive. Humanity cannot permit known carbon deposits to be consumed in their entirety. The effective limit on fossil fuel combustion will in all probability be reached sooner than the absolute limit of physical reserves. Should fossil energy consumption exceed the maximum tolerable load on the global ecosystem, the damage will be irrevocable and beyond the capacity of politics and economics to handle: dramatic cooling or warming of continents, floods or destructive storms and hurricanes.

But even if the crisis point is reached later than expected, the curves of declining reserves (of fossil fuels in particular) and rising demand will inevitably intersect. Economic chaos will have become unavoidable even before this happens. It will not just affect those countries least able to bear higher energy prices, primarily those in the developing world. Established industrial countries will also see their markets distorted and undermined, energy prices will rise beyond consumers' ability to pay, unemployment will go through the roof and there will be political tension and violence. And if demand eventually does outstrip supply, the result threatens to be the greatest bloodbath in human history – a hell on Earth, in which rational, organized action will scarcely be possible.

This apocalyptic scenario is unfortunately not unfounded

speculation. Indeed, factors promoting increased consumption make it all the more probable. The global population is predicted, for instance, to climb from 5.3 billion in 1990 to over 8 billion by 2010 alone, and it is also expected that urban populations will expand across the board. In 1990, 75 per cent of North Americans lived in cities; in 2025 the figure is expected to be 80 per cent. Over the same period, the percentages are expected to increase from 80 to 85 per cent in Western Europe, from 65 to 75 per cent in Eastern Europe, from 70 to 84 per cent in Latin America, from 33 to 55 per cent in Africa, from 32 to 54 per cent in China, and from 28 to 48 per cent in south and southeast Asia, in line with these countries' efforts to industrialize. These latter statistics for developing countries are particularly alarming. Urbanization implies a massive increase in energy and resource consumption. In the face of these figures, the 'weightless economy' can be seen for the ephemeral bubble that it is. Growth rates are more likely to be higher than lower, which means that the global economy will be exhausting fossil fuel reserves more quickly rather than more slowly. As conventional oil and gas reserves will probably be exhausted between 2030 and 2040, this period, as depicted in Figure 3.1, will also see the death agonies of the fossil global economy. Humanity will then be embroiled in a historically unparalleled fight for survival; and will lose, if the fossil global economy prosecutes its orgy of consumption to the bitter end. If humanity is driven to this intersection between supply and demand, the result threatens to be the most brutal military conflict in human history – truly the war to end all wars.

The true believers of the fossil-fuel economy have closed their eyes to all this. Their own reserve forecasts collide with their own predicted growth rates. There has not even been any attempt to calculate how much faster a fossil energy source will be exhausted once it has to replace an exhausted source. The chains of energy supply seem also to be fetters on thought. In this can't-see-won't-see culture, few have dared draw the logical conclusion that there must be a massive shift towards renewable resources – and those who do speak up are cautious and timid. The soi-disant realists of the established resource

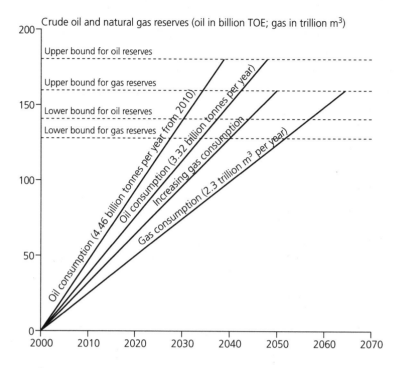

Figure 3.1 *Estimated duration of crude oil and natural gas reserves*

industries talk up the last remnants under the earth while disparaging the potential of renewable resources. Fossil fuel utopias are thought realistic, whereas comprehensive scenarios for solar energy are deemed unrealistic. Finite resources can be exploited forever; inexhaustible resources are of limited use. This has been the great contradiction of the 20th century. Without change, it will also become the tragedy of this new century.

Arming for the resource conflict

The only reliable means of avoiding the global resource conflict is to go cold turkey on fossil energy as soon as possible. Instead, however, the transport networks for fossil resources are being expanded to allow large and faster flows, and countries are beginning to arm for the coming conflict.

Geopolitics, long consigned to the past, is now experiencing a renaissance. The basic values of civilization, such as support for democracy and human rights, are usually translated into action only when this happens to suit other ambitions – witness the Gulf War. In times gone by, rulers have sought to expand their territory, whether to feed their own megalomania, to erect a defensive perimeter or to acquire additional space for settlement. Nowadays, such ambitions of territorial annexation are hampered by modern international law and global public opinion. In any case, the major industrial countries, because of their highly developed agricultural technology and stagnant population dynamics, have no reason to seek territorial expansion. High-speed transport and communication networks and the structure of global corporations also render it unnecessary to place new markets under direct political control.

The classical modus operandi of geopolitical resource security was territorial colonialism. Colonies, however, required political administration and standing armies. With the rise of domestic liberation movements backed by international support and spurred by the ascendancy of the democratic ideal in the 20th century, colonialism became inefficient and inopportune. Indian independence and the successful communist revolution in China immediately after the second world war sounded its final death knells.

Modern colonialism in the form of global capitalism is vastly more effective, but can only be achieved with the aid of the resource extraction companies. There is no need to assume political responsibility for the dependent territory, so the costs of administration and policing can be saved. The only requirement is the need to ensure that the governments of resource-exporting countries do not interrupt the flow of resources to the industrialized world. All insurrections have so far been crushed, some quickly, others slowly.[14] The goal has always been secure access to reserves and the lowest possible prices for resources. When governments of resource-exporting countries nationalized the extraction industries in order to increase their share of the profits, sanctions were and continue to be used to force cooperation, up to and including economic embargo. The most up-to-date method is to make International Monetary Fund (IMF) loans contingent on deregulation and privatization of industry. Even the OPEC cartel, which provoked a worldwide crisis between 1973 and 1982 with a threefold increase in oil prices, is scarcely able to cause a second, similar disturbance: the oil-exporting countries have bought large company shareholdings in the industrialized countries, and thus have no interest in bringing about economic downturn in the industrialized world. In any case, the industrialized countries are skilled exploiters of divergent interests and political conflict between the oil-exporting countries, judiciously applying the age-old principle of divide and conquer.

By contrast, attempts by resource-exporting countries to bolster their revenues within the framework of UNCTAD were never successful. The resource-exporters were too divergent and too short-sighted in their interests, and too pliant in the face of superficial offers of developmental aid; the élites of some countries were too strongly tied to the former colonial powers, or governments were too heavily swayed by the resource corporations. Western influence on international organizations was also too great for the attempt to create a counterweight – a kind of 'trade union' of resource exporters – ever to have worked.

The tactics employed by the industrialized countries to further their resource interests were and continue to be extremely dubious, including stirring up, or at least exploiting, disputes and wars between resource-exporting countries. In the 1980s, for example, Saddam Hussein served as the industrialized countries' weapon against the Islamic fundamentalism of Iran: he was allowed to conduct a brutal war before he himself trespassed on the oil interests of his Western patrons. Western industrialized countries have also not been above supporting ruinous kleptocratic regimes just as long as they helped keep the resources flowing. Witness, for instance, the fall of Mobutu in Zaire, whose successor Kabila was helped to power by the USA because he promised to better serve their resource interests. Kabila's first move was to redistribute the mining rights in the conquered areas. Behind the armed resistance movement which has since risen against him stand yet more resource companies. Alternatively, warlords have been financed to neutralize other warlords who threatened to impede resource flows: although Somalia still lacks a functional state, the oil wells of US firms are once again flowing. Even fundamentalist religious fighters like the Taliban in Afghanistan have before now been supported by the USA. Support for the Taliban in particular continued even after the Russian troops had long since been withdrawn from Afghanistan and the communist regime had been swept away – despite the USA's misgivings about Islamic fundamentalism in, for example, Iran. The only possible explanation for this was that the Taliban had undertaken to guarantee the passage of oil from the Transcaucasian states for US oil firms.[15] The tragic events of 11 September 2001 are a sad illustration of the folly and short-sightedness of this strategy. Mercenary companies equipped with the latest weaponry and aircraft are also bringing a new directness to the question of resource security. One example is the Strategic Resource Corporation, owned by the South African company Barlow, which has lavish headquarters in London and is also active in the resources business. The Strategic Resource Corporation undertakes security operations for mining and oil installations and hires its services out to governments and opposition forces in Africa for military engagements.[16]

Only when the resource-exporting countries have become ungovernable, when even the puppet governments of the resource interests of the global economic powers have lost their authority and can no longer be replaced by better-functioning regimes, do the armies of the West return. This is usually justified by reference to humanitarian intervention, and many regions are indeed anarchic. There is even a UN mandate for 'humanitarian actions'. Nevertheless, closer examination shows that the decision over where such interventions are or are not made has far more to do with resource interests, especially in Africa. The Rome Declaration on Peace and Cooperation, agreed at the meeting of the North Atlantic Council in Rome on 7–8 November 1991, which formulated the 'new strategic concept' for the NATO alliance following the collapse of the Warsaw Pact, contains the following passage: 'Our Strategic

Concept underlines that Alliance security must take account of the global context. It points out risks of a wider nature, including proliferation of weapons of mass destruction, disruption of the flow of vital resources and actions of terrorism and sabotage, which can affect Alliance security interests.' In response, it was decided to create 'ground, air and sea immediate and rapid reaction elements able to respond to a wide range of eventualities, many of which are unforeseeable'.

This new concept of geopolitics is none other than the old politics of global resources. The representatives of the major industrial powers are becoming ever more blatant in the way they subordinate the fates of entire regions to their own resource interests, sacrificing other maxims of international politics as they go. When the supply of resources is threatened, it is the entire economy – and thus naked self-interest – which is at stake. This can be seen most clearly in respect of the treatment of the former USSR and of Russia in particular. It is not the stabilization of Russia and support for democratic institutions which are at the forefront of political strategies, nor is it the attempt to avert the alarming collapse of a state which is still a nuclear power. Here, too, it is control of resource reserves through the neutralization of Russian influence which is the real priority. The goal is to direct the flow of resources from the newly independent states around the Caspian Sea, from Azerbaijan and Kazakhstan to Turkmenistan, along the channels of Western interests. US involvement in Central Asia has historically been solely concerned with resources and it remains, as Zbigniew Brzezinski emphasizes in his book *The Grand Chessboard*, of 'primary importance' to the 'global supremacy' of the USA.[17] Of course, since 11 September 2001 there has also been a security dimension to US interest in the region. His arguments give the lie to apologists who decry the coming resource bottlenecks as delusion or merely a question of technology and mathematics.

Such policies are now also practised beyond the Middle East and Africa – and they are employed to secure other resources than just fossil fuels. 'Apocalypse soon ... in Minerals' was the alarming headline for an article in the US security journal *Defense Week* as long ago as 1980.[18] For resource-rich South Africa, this fear even had a positive effect. After long years of Western support for the apartheid regime, following a change of tack by the then US President Carter, the Western powers finally distanced themselves from their racist friends as the political fallout, in particular in terms of the growing influence of the USSR in post-colonial Africa, threatened to grow too large. But it was not just the ideologically charged East–West conflict which dominated US foreign policy in this period, as the majority of studies and commentaries assert. A substantial, often even the dominant, role was always played by geostrategic resource interests, even in the East–West conflict itself. This was openly debated in the USA, whereas the Europeans chose to close their eyes in post-colonial shame. The term 'strategic resources' has a double meaning: any country with important resources buried in its soil is part of the strategic sphere of interest of US foreign policy.[19] The political interest of the USA in the Caucasus and in central Asia has until recently stemmed primarily from the same motivation. In *Global Monopoly*,[20] Gernot Erler documents a new 'politics of containment' with a 'decidedly anti-Russian slant', the aim of which is to permanently exclude all Russian influence from these regions. It is in these Eurasian territories, however, that the interests of Russia, the two most populous and resource-hungry countries India and China, the Islamic hemisphere and the EU intersect. The danger is that they will become the primary theatre in the battle for the world's dwindling resources.[21]

In April 1999, on the occasion of the North Atlantic Council meeting to mark the 50th anniversary of NATO, the strategy initiated in Rome in 1991 was reiterated with the unambiguous intention of turning NATO into a hegemonic alliance for the safeguarding of Western values and interests, whereby these latter two terms were taken as synonyms. Rapid reaction forces had in the meantime been established in almost all NATO states, and these were to be further expanded. NATO now explicitly reserves the right to intervene militarily in other regions. If necessary, it will do so even without a UN mandate. Above all, it is mounting a targeted campaign to increase its membership.

After Eastern Europe, Asia is the next target for NATO

expansion. The oil- and gas-exporting countries of the Caucasus and Transcaucasia have already become so-called 'NATO partner countries', from Azerbaijan to Kazakhstan, Turkmenistan, Kyrgyzstan, Uzbekistan and Tadjikistan. None of these countries is democratic, all being ruled by clans and oligarchies, yet they are nevertheless being treated as candidates for NATO membership. In this, NATO's only motive is the resources that these countries possess. These reserves lie both in the midst of former Soviet territory and on the doorsteps of the emergent giants of China and India, whose burgeoning populations have the greatest need for additional resources. An attempt by NATO to secure privileged access to the last substantial reserves of fossil fuels remaining untapped would be a slap in the face to both Russia and to 2 billion Indian and Chinese citizens. Were these three states to take the logical response of forming a triple alliance, the result would be a new East–West conflict – whose Asian focus would also make it a North–South conflict – accompanied by a renewed conventional and nuclear arms race.

The magnitude of the risk shows up starkly against the backdrop of China's and India's growing thirst for energy. In 1976, the demand of Chinese power stations for fossil fuels was still only 175.8 million TOE. By 1996 this had risen to 877.8 million TOE, which is a fivefold increase in only 20 years. Demand for oil alone, which China was long able to meet from domestic supply, rose by 85 per cent between 1986 and 1996. This almost twofold increase in ten years has been accompanied by growing Chinese demand for oil imports. The demand of the Indian electricity industry for fossil fuels grew from 57 to 357 million TOE between 1976 and 1996, a sixfold increase. Over the same period, fossil fuel consumption in power stations grew five and a half times from 300 million to 1.66 billion TOE. Crude oil consumption in the rest of Asia outside of Japan, Korea and the Asian republics of the former USSR has tripled since 1976, and doubled between 1986 and 1996[22] (see Table 3.2).

Besides the threat of environmental catastrophe, the conflict between the needs of the Asian region and the growing and – despite the enormous growth in Asian demand – dispro-

Table 3.2 Growth rates for fossil energy use in Asia, in millions of tonnes

	1976 Annual quantity	1981 Annual quantity	Increase on 1976 (%)	1986 Annual quantity	Increase on 1981 (%)	1991 Annual quantity	Increase on 1986 (%)	1996 Annual quantity	Increase on 1991 (%)
China									
Fossil fuel consumption in power stations	175.8	253.28	44.0	355.0	40.2	552.46	55.6	877.77	58.9
Oil consumption	77.7	83.1	6.9	97.2	17.0	119.6	23.0	172.6	44.3
India									
Fossil fuel consumption in power stations	57.51	89.89	56.3	142.39	58.4	237.29	66.6	357.54	50.7
Oil consumption	24.86	35.55	43.0	45.7	28.6	59.96	31.2	82.19	37.1
Asia (total)*									
Fossil fuel consumption in power stations	300.68	468.1	55.7	812.91	73.7	1065.2	31.0	1657.57	55.6
Oil consumption	161.3	219.7	36.2	255.1	16.1	339.9	33.2	481.1	41.5

* Not including the OECD countries Japan, South Korea and the Asian republics of the former USSR.
Source: International Energy Agency

portionate energy imports of the leading industrialized nations raise the spectre of 'cold' trade wars and 'hot' shooting wars to come.

Yet rather than looking to renewable energy, the geopolitics of global resources instead revolves around checkmating potential competitors or playing them off against each other. This strategy takes two forms. On the one hand, there is the unrelenting campaign to impose a world market as the global economic order, in order to keep all doors open for unimpeded imports. In a world market, the transnational corporations' large capital stock for investment, purchasing power, technological edge and dominant influence on international institutions of the Western industrialized countries gains them the lion's share of global resources. On the other hand, these nations are also enhancing the superiority of their military technology and their ability to conduct global operations, in order to be able to use military force, if need be, to put competitors in their place.

Many experts are of the opinion that competition for access to vital water reserves is most likely to escalate into outright war. Tensions are greatest where two countries depend on the same major rivers for their water. Whoever controls the source of the river is in a position to crowd out downstream countries. Thus Sudan has the advantage over Egypt along the Nile, and Turkey over Iraq along the Euphrates.[23] Albeit these conflicts are a matter of life and death to the countries concerned, they nevertheless remain restricted to particular regions. The competition for fossil fuel resources, on the other hand, has global dimensions, even though the visible effects have to date been more regional. Dwindling reserves will almost by necessity force a radical shake-up of the global political scene in the decades to come. Even if it never comes to war, the arms race has already begun.

The connection between nuclear proliferation within the Islamic–Indian sphere and the competition for resources is impossible to ignore. The leading industrialized nations have been working together to isolate Iran following the fall of the Shah in 1979 just as effectively as they cooperated in the near-execution of Iraq in 1991. Both of these were clearly punitive actions taken against states who dared upset the resource status quo. The subsequent occasional missile attacks on Iraq are intended to frustrate the country's aspirations to become a nuclear power. The knock-on effect, however, has been to push other nations in the global competition for resources to accelerate their own nuclear weapons programmes. Once these countries are established nuclear powers, the USA will no longer be able to push them around as it did Iraq: this is the lesson that was taken from the Gulf War. Iran's nuclear programme is founded on this calculation. Albeit the nuclear arms race between India and Pakistan that produced the spectacular weapons tests of 1998 may be primarily a product of the old enmity between these two countries, the overarching motivation is the greater bargaining power that nuclear weapons bring. Chinese politicians justify their uncompromising stance on nuclear weapons in explicit terms of the need for a position of strength in the light of China's rapidly growing dependence on imports.

Total Western defence expenditure has by no means fallen by as much as the end of the cold war would have suggested, despite the almost universally parlous state of public finances. According to calculations performed by the International Institute for Strategic Studies in London, defence expenditure in 1986 ran to $585 billion within NATO, and $343 billion in the USSR. In 1997, NATO expenditure was still $454 billion, whereas the Russian defence budget (albeit pertaining to a smaller area than the former USSR) had shrunk to $64 billion. If NATO accounted for 48 per cent of global defence expenditure in 1985, this figure is now 57 per cent – even without the numerous NATO 'partner countries'. Terrorist threats, while undeniably real, do not wholly justify this expenditure, as the fight against terrorism cannot be won by conventional warfare alone. Russia's internal crisis and the fragmentation of its army make a Russian threat unlikely. The sole remaining legitimization is the power to intervene in regional crises, crises which increasingly have the character of resource conflicts. The global political likelihood is that those countries with growing demand for fossil fuels will in future want to claim a larger slice of the dwindling cake, and build alliances against the US–European–

Japanese axis that currently controls access to global reserves. Possible alliances include not just China with Russia or India, but perhaps also Russia with Iran, or even with a Turkey spurned by the EU; or possibly Turkey with Iran, Pakistan and China. Another possibility is that Japan might sell out and ally itself with China, and possibly with India and Indonesia as well, with a view to Australian resources. The USA's emphatic endorsement of Turkey's candidature for EU membership (despite Turkey's well-documented human rights abuses) also has the resource-strategic objective of forestalling any other alliances. It is no coincidence that south, southeast and central Asia are the only regions where defence budgets have been sharply increased: from $120.6 billion to $160.8 billion between 1985 and 1997, measured at 1997 exchange rates.

Resource reserves, gunboat diplomacy and the moral bankruptcy of society

Be it Russia or Somalia, Indonesia or Mexico, Congo or Sri Lanka, Yugoslavia or Algeria, Angola or Georgia, Nigeria or Afghanistan, Rwanda or Uzbekistan, instances of political collapse and bloody conflict are becoming ever more common. These ethnically, religiously or nationalistically motivated conflicts offer a taste of what is to come, as

- the already highly unequal distribution of global fossil fuel reserves will inevitably deteriorate further as stocks are depleted; and
- the increasing environmental damage wrought by fossil fuel consumption, alongside potential nuclear accidents, will devastate the livelihoods of ever more people.

The resulting shortages of land and sustenance will be the spark for escalating violence and bloody excesses, wherein the familiar mechanisms of anthropogenic selection will see their most tyrannical and remorseless application yet. What may at first sight appear to be an ethnic or religious conflict requiring the intervention of the enlightened guardians of law and order and human rights, is in reality the resource-centric self-interest of those self-same self-appointed guardians.

Coarsened and brutalized relationships both within and between states and the likely disintegration of state structures loom large on the horizon of the 21st century. Some countries, like Somalia, fall into anarchy; others, like the USSR, with Russia possibly soon to follow, collapse into ever smaller fragments. Next on the list could be Indonesia, China or India. This disintegration cannot be always counted on to pass off as uneventfully – without threat to world peace – as has so far been the case in the former USSR. Whether even the EU itself will survive the economic turmoil when resource shortages begin to bite is also more than debatable.

The existence of some sort of global governance has made it possible on numerous occasions to dampen down conflicts which those involved have no longer been able or willing to settle themselves. Outside authorities can sometimes be a force for moderation, and can help to construct a new social order. This remains a possibility, for as long as gulfs can still be bridged the conflict remains soluble, so long as the helping hands are truly interested in defending common values – protecting people and achieving an equitable outcome. The question of resources, however, is a raw nerve for the large industrialized nations, and where there is dispute, the global players of the political and economic world are single-minded and unscrupulous in the pursuit of their own interests. 'Global governance' is in this case doomed to fail, because when it is their resource interests that are at stake, the dominant powers not only lack the necessary credibility to be accepted as impartial arbitrators, they also have no real desire to achieve an equitable settlement.

All the giants of the political stage have at least indirectly contributed to the ecological destruction of human habitats and the social constraints that result from unequal access to resources. Despite the historic scale of their global power, they have so far proved incapable of global responsibility. They rely first and foremost on the economic agents that have arisen to manage food, energy and resource supplies. Not only have they proved themselves able to supply society with cheap food, but in the form of corporate empires, these economic global players have also proved superior to all other business forms in the global marketplace, both in efficiency and range of products. The result is a determination not to upset the applecart, on the basis that these structures are the only way to secure economic livelihood.

On the one hand there are the transnational corporations, whose scope is continually being widened not just by themselves, but also by political agreement. It is not just the WTO that draws no distinction between environmentally friendly and environmentally harmful products, or between finite and renewable resources; so does the increasing tendency of both national governments and international agreements to smooth the way for transnational companies. There is scarcely an investment made by these companies which does not attract subsidies, whether the publicly funded provision of business parks and infrastructure, several years of tax breaks or direct investment grants. There are few large mergers which are not welcomed or even actively supported by the host country. As governments are no longer in a position to shape their own futures, they hope that transnational corporations will do it for them. This is the thinking that lay behind the proposed Multinational Agreement on Investment (MAI). The effect of this agreement, which suffered at least a temporary setback in the summer of 1998,[24] would have been to shield foreign direct investment – which in the case of transnational companies covers effectively all investment – from subsequent additional political obligations with unknown cost implications, essentially new legislation on social, tax or environmental issues. Governments would have become liable for any additional business costs that might result from such legislation. This would have effectively exempted all corporate empires from domestic legislation, turning them into international institutions with unlimited power but no political or social liability.

On the other hand, there are the global environmental and the social consequences, whose effect is felt nowhere more keenly than in the rural regions of the developing world, the incubator for the greatest sociocultural debacle in world history. Three billion people, half the world's population, are sustained by agriculture in these regions. As agriculture must increasingly look towards the global market, these people will become ever more dependent on the global food-processing corporations, whose industrial logic leads them to move towards mass production in large-scale agribusiness. Their monopoly on purchasing gives them the power – through market forces – to reshape business structures more imperiously than ever.[25] Assuming this standard model of agricultural modernization under world market conditions, the likely consequence is that two of these three billions will lose their livelihood without any prospect of anything to replace it. At the turn of the century, the Chinese sociologist Feng Lenrui estimates that there will be 170 million unemployed agricultural workers in China – one quarter of the entire workforce.[26]

The consequence: either we embark on the road towards a solar resource base and correspondingly towards agricultural structures that do not follow the industrial model of the global market, or we abandon the values of the modern age because the global economic 'reality' seemingly admits of no other world. From the ideal of international equity to the maintenance of the natural basis for life; from the ideal of a social balance within states to the ideal of the constitutional democracy – all subordinated to the need for unimpeded access to resources and their optimal economic deployment. Anything may happen – from apathetic acceptance of ruinous trends to the closed eyes and ears of the leisure society, from turmoil and conflict in disintegrating societies to the blatant use of military force in support of business interests, from the decay of order within nations to the collapse of hard-won but underdeveloped international agreements – to the point where 'chaos reigns' (Samir Amin).[27]

Because nothing is so crucial to survival as a secure resource base, where it is under threat for part or all of humanity there is willingness to commit acts of extreme ruthlessness and barbarity. Ruthlessness can already be seen; barbarity will follow as the threat to resources becomes more acute. It is hard to imagine what faces the world if the opportunity to switch to solar resources is not seized – including a return to agricultural methods which preserve the fertility of limited agricultural land and the cleanliness of limited water supplies.

If the only policy options open were continued consumption or abstention, humanity would stand little chance of not tearing itself apart. As plausible as it may sound, forgoing consumption may well prove to be an unworkable solution. There are no satisfactory answers to the questions of who should enforce it on whom, to whose benefit and at whose cost. Resource needs are too basic and access too unequal for consensus to be achieved at national level, let alone on a global scale. Many need and want more, but no-one is prepared to take less.

Only a solar resource base offers any escape. The apologists of the fossil global economy justify their failure to make even half-hearted progress along this road with an equally tired fossil of an argument: in a world of global competition, the 'luxury' of concern for the environment must be earned through further conventional economic growth. This economic philosophy is in reality a necrosophy – the wisdom of death. Its absurd consequence is that the price for safeguarding the environment is the freedom to continue – for how long? – damaging it. This 'wisdom' has dominated our culture for too long, and we can no longer afford to heed it.

CHAPTER **4**

The distorting effects
of fossil supply chains

THE DISTORTING EFFECTS of fossil supply chains can be seen most clearly at the extreme ends of the sociocultural spectrum – in the cities of the industrialized countries and the rural regions of the developing world. It was supplies of fossil energy that first made the megacities of the modern age possible; now, as supplies of fossil energy near exhaustion, those megacities are threatened with collapse.

The rise to dominance of fossil energy and the emergence of a global energy monoculture had severe knock-on effects on rural regions. Developing countries are experiencing today what the industrialized countries lived through in the past. Despite abundant solar energy, migration from the land to the slums of the cities continues to increase because people see no way of earning a living from agriculture. One reason for this is the lack of effective energy systems, without which neither agricultural structures nor new business initiatives can develop. Plenty of sun, but no energy for economic development: that is the grotesque situation of the developing countries, rather as if the crew of a ship had run out of drinking water, and its captain could only lament, 'Water, water all around, but not a drop to drink.' An on-board desalination plant would have prevented this life-threatening problem from arising. Similarly, the developing world urgently needs a solar energy industry to ensure future prosperity.

Patterns of settlement and social organization have always been heavily influenced by the nature and available reserves of energy supplies.[1] As we enter the twilight of the fossil energy age it is all the more important that we should tackle the grave problems facing rural and urban areas alike, and consider how they might develop in a post-fossil-fuel age. The energy question currently does not figure in discussions of the future of town and country, such as the Habitat conferences initiated by the United Nations to examine the future prospects of urban lifestyles. This omission must be rectified.[2]

The rise and fall of the fossil city

While there were still no technologies available to harness energy efficiently, no fast and capacious means of transport and no transport infrastructure, settlements had to be located directly where energy and food were produced. In pre-industrial times, a town needed an area of fields and woodland for its energy needs between 40 and 100 times as large as the actual inhabited area, depending on the local climate and soil quality. Dearth crises were unlikely as long as this natural growth limit was respected. The structure of the economy was not conducive to amassing wealth, but its stability was assured as long as peace prevailed.[3] There were differences in living standards, resulting from natural conditions or from varying levels of technological and cultural development, but these differences were smaller than the global disparities of the fossil industrial age, especially following the industrial concentration and monopolization of the fossil energy system in the first half of the 20th century, and its ultimate globalization in the second half.

At the height of the Roman Empire in the 1st century CE, the inhabitants of Rome numbered perhaps half a million. It was only with the availability of concentrated energy supplies and the constantly improving technology of the industrial revolution that urban expansion and the shift in settlement patterns from country to town could take place. The ability to transport food and energy into the towns and cities from ever more distant regions, with ever increasing technological ease and at ever lower cost, was crucial to this development. In 1800, there was one city on the entire planet with more than 1 million inhabitants. In 1900 this had risen to 13, by 1990 it had reached 300.[4] The urban industrial centres developed first in regions with large coal deposits, then later along the main lines and flows of the increasingly concentrated energy supply networks. In so doing, they also became centres for the energy-intensive service industries. The faster the fossil megacities grew, the more inorganic their sprawling growth.

The fossil city grew only slowly in the first stages, as not all the technologies for harnessing and transporting fossil energy were yet in place. Initially, the only option was to ship large quantities of primary energy to ports on the coast or along rivers. Then came the railway, which opened up the hinterland for large-scale urban growth far from the major navigable waterways. Electricity cables followed, enabling energy to be transported even faster across an even wider area. At the same time, technologies for energy use underwent rapid development. New technologies were coming on-stream all the time; the infrastructure of energy supply and transport was expanding rapidly and energy was flowing into the industrial metropolises as never before. The greatest and fastest growth spurts came from the construction of national power grids and the mass production of automobiles, which increasingly came to shape the development of towns and cities. Nevertheless, there was, in the initial stages at least, still scope for gradual adaptation to industrial developments and thus also opportunities to shape these developments as they occurred.

Students of urban sociology today almost all expect continued migration towards ever larger megacities, thinking the trend towards global sociocultural homogeneity to be irreversible. But in accepting this development as predestined and immutable, they fail to see just how fragile the energy base of these megacities is becoming — and in the developing countries indeed has already become. It does not usually even cross their minds that a change in the course of development is not only desirable, but also possible and desperately needed. Great effort is expended on cataloguing and managing individual problems, yet it has not been possible to prevent these problems from growing in size and spilling out over attempts to contain them. Continued urban expansion can only prolong our dependency on fossil fuels, yet few realize that our present energy system can only be a historical interlude. Only if the fossil energy system, including nuclear power, were to be replaced by another centralized system would a continuing shift towards mega-urban structures be a theoretical possibility, albeit at the price of the continued destruction of rural communities.

The centralized energy system is also responsible for the creeping homogenization and impoverishment of urban life. When the fossil energy system first gave rise to the industrial centres, they attracted people like magnets. The new cities promised work and easy wealth. The only force acting to counteract or moderate the exploitation and mass misery of the early industrial age, as documented in lurid tones by the first modern social critics, was the growing influence of the organized socialist movement. Cities have shaped our model of civilization: industrial labour, a multiplicity of educational, career, leisure and cultural opportunities, mass media. The growth in capital stock allowed construction on a mammoth scale: buildings for factories, gasworks and waterworks (which now house luxurious apartments); ironwork, from roofing for halls to stations and bridges, complete with latticework and artistic mouldings.[5] The only way was up – until, that is, the industrial cities themselves fell into the growth trap. Their industrial foundation vanished, the internal structures it had supported collapsed and environmental degradation gained the upper hand.

The role of energy in the decline of urban diversity

While the scarcity of energy was still a daily reality in almost all the world's civilizations, living quarters necessarily had to be built from locally available materials, and architecture had to take the prevailing climate and ecological situation into account, making the best use of sun or shade, insulation or cooling. Trees were used as shelter from the wind, slopes for heating, windbreaks for cooling; buildings were circular to maximize energy efficiency, and wood, local stone or earth were used as building materials.[6] Across the world, in towns and villages alike, the result was locally coloured variety in building construction, style and materials – the 'evolution of solar architecture' (Behling).[7] The 20th century, however, marked a new departure in architectural history. Where supplies of oil, gas and construction material were easy to obtain, buildings soon lost their climatic and regional variations.

Plentiful supplies of energy and materials gave architects and town planners a completely free hand, unfettered by the restrictions of local climate, geography and ecology. And yet, ironically, the result has been global architectural uniformity. Natural cooling systems were redundant; the electricity grids provided abundant energy for air conditioning and refrigeration. There was no need to make use of natural sources of heat because it was no trouble to ship heating energy from halfway around the globe. Centralized energy and materials supplies presented a golden opportunity to reduce construction costs through mass production and standardization. This resulted in buildings which had ever fewer original features, quickly lost their distinct identity and which had ever shorter useful lives before needing repair or demolition. Whether Berlin or Rio, Paris or Athens, Sydney or Boston, the buildings of architectural modernity were uniform, interchangeable and often difficult to tell apart.

The planning philosophy of the fossil city was formulated in the 1940s by Le Corbusier (1887–1965) in his *Athens Charter*: division into separate functional zones for living, working, shopping, leisure and traffic.[8] Though it spoke to the problems of the fossil city, the Athens Charter did not seek to question the nature of the city itself. Residential areas were to be sheltered from traffic, but at the cost of a huge rise in traffic flows. Where the intention had been more a more functional urban space, the outcome was more complexity, greater loss of people's time and disrupted communications structures. The emblems of the fossil city are the disconnected spaces of its functional zones: industrial estates, shopping, sport, health, leisure and cultural centres. The result was a functionalism that put a premium on mobility, with ever more space allotted to streets and cars, disrupting the organic life of the various parts of the town. The sole force binding the city together was the dominant traffic corridors. These were also its greatest burdens.

The key element in this model of the city was waged employment in the industrial centres. Where this is lacking, the fossil megacity is faced with internal collapse. Grand visions of the future 'global city' are an attempt to paper over this trend. Pundits wax lyrical about metropolises which are home to the headquarters of the global players and a happy hunting ground for providers of consultancy, information and financial services, the advertising agencies, hotel chains and IT firms[9] – but how many of the world's cities can truly fill this role, given the global trend towards mega-mergers and ever fewer corporate headquarters, yet at the same time an ever greater urban population?

The changing world of work and the erosion of the megacity

The initial concentration of energy supply infrastructure in the cities made them privileged economic centres. As fossil energy systems have gone global, however, and networks have been expanded, and as industrial structures have changed across the world, this tie has been weakened. In many industries, the speed and ease with which headquarters can be relocated has grown, beyond even the ability of an increasingly mobile workforce to keep up.

This new corporate mobility is leaving behind the inhabitants of the megacities in their functionally separated residential areas whence they once swarmed out to go to work, to go shopping or for leisure activities. Without sufficient work and wages to support it, urban zoning becomes senseless, leading to ghettoization. Highly paid careers in professional services stand in stark contrast to the increasing numbers of low-skilled service jobs in restaurants, delivery services and cleaning services, usually with low wages paid by the hour. In this 'self-destructive yearning for the global city', as Hartmut Häußermann has described the eulogies for this model for urban 'modernization', the rising tide of the unemployed and the low-paid is accompanied by increasing dependency on state- county- or borough-level social support.[10] In consequence, urban tax revenues are falling. In turn, the financial difficulties of borough councils are forcing metropolitan authorities to privatize borough-level responsibilities and to commercialize what were previously cost-free or low-fee public services; access to public spaces becomes more difficult or even unaffordable for many; social stratification increases and rifts widen, leading to increased tension and criminality. The industrial city is doomed to fail. We are nearing the crisis point where fossil fuel reserves fall and prices rise beyond the means of the majority of city-dwellers. As the rot affects greater numbers of urban inhabitants, the first megacities of the established industrialized countries are already sinking inexorably to the level of the developing world. The optimistic visions of urban planners are turned to rubble. As a model for civilization, the fossil megacity has no future; and for most of the world's cities and their inhabitants, the 'global city' is no more than a cruel illusion.

How does the fossil energy system impact on this development? The urban population bears the direct and indirect costs of energy, supplies of fuel and electricity, heating and cooling systems and motor vehicles. In Germany, average annual per capita expenditure on energy is more than €2000 (over $1700). This figure does not represent the size of an individual energy bill, as it also incorporates the energy component of expenditure on services and the energy usage of business. If a city's entire energy needs are met from fossil fuels, then for a city of 1 million, that equates to the sum of €2 billion ($1.7 billion) abstracted from the city's economy every year. Including food expenditure of €1500 ($1300) per capita, or €1.5 billion ($1.3 billion) in total, this million-inhabitant city must pay a total of €3.6 billion ($3 billion) for its food and energy imports.

The running cost of fossil energy supplies has to be covered by value added in the urban economy. For a long time this worked. Value was created primarily within cities and there was sufficient work on offer for the urban population. However, as industry replaces human labour with machines or finds cheaper labour elsewhere, and as sufficiently well-paid jobs become increasingly scarce and urban unemployment rises to 20 per cent and beyond, the megacities are subjected to a process of wastage and impoverishment. The need to meet daily requirements for food and energy is universal, but in the cities, food and energy must also be financed and imported, which makes it more difficult to secure supplies. Destitute inner cities are the unhappy consequence.

The dilemma is stark, yet the response is obvious. The concept for the future is not the 'global city' but the 'solar city'.[11] Energy sources must find their way back into the city, and not just to make cities habitable again. 'The freedom of the city' might have been the motto in decades past, as the towns and cities of the newly industrializing countries promised a cornucopia of new opportunities for upward mobility. Now the free energy of the sun must be harnessed to liberate individuals and economies from their dependency, to break the shackles of monthly energy bills and render it easier to achieve self-sufficiency in basic needs: food, energy, living space and the opportunity to partake of cultural life. The 'solar city' strengthens the economy of the city by the quantity of renewable energy it produces.

This is an idea whose time has come. Even in the industrialized countries, increasing numbers of town- and city-dwellers are finding that they can reduce their living costs by producing some of their own food. 'Urban farming' is not solely a feature of developing world cities. There are cases where twice as many people make their living from inner-city market gardens as from employment at the minimum wage. There are probably millions in Russia who would have starved over the past few years were it not for their home-grown vegetables. Even in the USA in the 1980s, urban food production grew by 17 per cent.[12] Obviously that cannot be replicated in every city, as there is limited space available for crops. Nevertheless, this trend is an indication of how cities are becoming dependent on their inhabitants' ability to produce their own primary goods. It also highlights how the existence of agricultural enterprises in the surrounding area will become vital for the future development of cities.

There is much greater scope for energy self-sufficiency in the megacities than for self-sufficiency in food. It is entirely possible that in the long term, cities can meet all their energy needs from solar energy. This may not be enough to halt the decline of the city, but it is nevertheless an essential precondition for the regeneration of urban areas.

The fossil resource trap closes on the developing world

The model for developing world cities in the post-colonial era has, of course, been industrial development and the energy-intensive growth model. The result has been and continues to be a growth explosion with direct and immediate effects. Cities have not had time to adapt. An unprecedented wave of immigration has quickly overloaded their infrastructure. Towns have been 'opened up' with hurriedly erected and fast-decaying concrete blocks, connected by a welter of new streets and cabling; they have been engulfed by a ring of shanty towns and desperate slums, and their centres are permanently shrouded in smog. Cities such as Mexico City, São Paulo, Lima, Cairo, New Delhi, Mumbai (Bombay), Jakarta, Istanbul and Karachi, whose populations have long since passed the 10-million mark, are ample testimony to the hopelessness of fossil fuel civilization.

Most of the cities of the industrialized world have reached their growth boundary; overall, their populations are stagnant and, following the marginalization of the agricultural sector, only a small fraction of the population lives in rural areas. The megacities of the developing world, however, are faced with an interminable wave of migration which they are helpless to confront. The majority of the population of the developing world still lives on the land: 80.4 per cent in China, 77 per cent in India, 75 per cent across the rest of Asia as a whole, 73 per cent in sub-Saharan Africa. Vast numbers of people stand waiting at the gates of already hopelessly overloaded cities. The very term 'least developed countries' (LDCs) suggests that high rural populations are an indication of lack of development. Particularly high rural populations are found in Burundi (95.7 per cent), Rwanda (94.3 per cent), Burkina Faso (91.5 per cent), Uganda (91.2 per cent), Malawi (90.9 per cent), Ethiopia (89.5 per cent), Niger (87.5 per cent), Eritrea (86.5 per cent), Tanzania (82.2 per cent) and Kenya (83.9 per cent) in Africa, and in Nepal (93.5 per cent), Bangladesh (88.7 per cent), Cambodia (87.6 per cent) and Laos (86.6 per cent) in Asia.[13] The implicit message is to 'get out while the going's good' – as if the purgatory of the city slums were an advance on rural destitution.

The ultimate causes of this migration are poor management or complete absence of agricultural development. In particular:

- agricultural 'modernization' has robbed smallholders of the means to make a living; and
- the rural population either cannot afford the commercial energy supplies necessary for economic and cultural development, or have no means of accessing them because governments have concentrated their efforts on supplying the cities.

Rural depopulation and the catastrophic growth of cities are the visible consequences of the energy trap in which the developing world finds itself.[14] Hardest hit is the poorest continent, Africa, which Axelle Kabou describes in her book *Et si l'Afrique refusait le développement?* (What if Africa says no to development aid?) as 'both under-developed and under-analysed'.[15] The only way out of this cul-de-sac is to use renewable energy to provide electricity and fuel for the economic development of agriculture, the trades and small businesses in rural regions.

Academic investigations into the causes of under-development and outbursts of violence in the developing world give no consideration to this fundamental energy crisis, which only demonstrates their intellectual poverty. One example of this blinkered attitude is a publication by Stiftung Wissenschaft und Politik (Foundation for Science and Politics), the official foreign-policy think-tank for the German federal government, which examines the question of conflict prevention and resolution in Africa: not one word on the energy crisis, which is a direct cause of territorial conflict.[16] In the late 1950s, the American Walt Rostow documented five stages of economic growth whose order of occurrence is constant and unchanging. Starting from traditional agrarian society, they lead through the first industrial activities to a general industrial boom. This economic activity is then transformed into market economic structures, culminating in an affluent urban society.[17] The social and cultural consequences resulting from the imposition of this Industrial Revolution-inspired developmental model on most developing world countries have been documented on numerous occasions. Attempts to build a developing-world socialism that would tie into local cultural heritage were also failures, as they lacked an adequate answer to the basic economic needs of developing countries. There was no concept of sustainable economic development that could have taken the place of the stagnating rural subsistence economy. Developing-world socialism was also founded on the industrialization model, but used centralized economic planning to achieve it. Agricultural smallholdings were collectivized, or attempts to preserve them failed to secure the provision of the accessible and affordable supplies of energy needed to run motors, agricultural machinery and industrial plant. In consequence, the productivity of smallholders and skilled tradesmen fell in comparison to large-scale industry, and they were pushed ever further out to the margins of society.

The alternative would have been – and still is, more than ever – to introduce autonomous, locally based energy systems. As the history of solar energy illustrates,[18] the necessary technology – from small-scale hydro to small wind turbines, from biogas plants to wood gasification – has been available for a long time, yet only a few countries have made even partial use of it. The millions of biogas plants that Chinese smallholders have built for themselves are one exception,[19] but even these are used only for cooking and heating, not for generating electricity or for producing fuel to run labour-saving machinery. The energy strategies of developing countries followed what was regarded as the progressive model of centralized fuel and electricity supplies. 'Modern' centralized systems, however, cut economic development adrift from its sociocultural basis. Ninety-seven per cent of Tanzania's generating capacity, for example, is available only in cities. Distribution grids were – and of economic necessity had to be – restricted to urban centres as long as electricity generation relied on central power stations.[20] Lesotho produces 93 per cent of its electricity in large hydropower stations, a situation that fosters economic and social trends which are foreign to the structure of the country; only 7 per cent comes from more appropriate small-scale hydro plants.[21]

The World Bank (and in its wake other development banks and respective national strategies for development) has been deliberately pushing this one-size-fits-all strategy for decades, closely paralleling the interests of the industrialized countries and the fossil resource corporations. This has been documented in many critical analyses of the World Bank's activities. Of the $292 million the World Bank pumped into projects in Brazil between 1952 and 1963, $264 million alone was spent on electrification with central power stations. The energy needs of the resource corporations have always soaked up a large proportion of developmental lending, in order to guarantee supplies of resources from the developing world to meet the needs of the West. Many hydropower dams were erected solely to supply cheap electricity for mining activities and ore-processing plants, often with fateful consequences for the environment. Around half of all aluminium smelters draw their power from such sources.[22] Developing countries were deliberately pushed into dependency on the crude-oil giants, for example by providing finance for roads rather than railways, or by refusing finance for countries to develop their own oil extraction and refining industries. The existence of independent capacity would have run contrary to the market interests of the oil corporations.[23] Equally, the oil companies exerted massive influence on the World Bank to finance factories for the production of fertilizers. A solid 58 per cent of all lending in 1979 was for such factories. 'Commercial farming' was promoted ahead of smallholders, entailing greater use of large agricultural plant and oil-derived pesticides.[24] In each case, the rationale for these projects was the low investment cost by comparison with the output of energy, resources and foodstuffs achieved. World Bank executives never questioned whether these projects were of benefit to the trade balances and the sociocultural development of developing countries, or whether they were in fact counterproductive.

More recent analyses have shown that this World Bank policy was not confined solely to the 1950s, 1960s and 1970s.

According to a joint study by US and European non-governmental organizations in 1997, since the signing of the World Climate Convention in 1992 the World Bank has financed further fossil energy plants, thus increasing rather than reducing emissions of greenhouse gases. Lending programmes for renewable energy have since at least got off the ground, but the World Bank is still ploughing much higher sums into supporting investments in new oil and gas fields, coal mining and fossil fuel power stations. The Bank also supports foreign investment in, and takeovers of, energy companies in the developing world, and by extension is facilitating the industrial concentration of world energy supply chains. Ninety per cent of projects benefit the energy companies of the seven largest industrial nations; only 5 per cent of the energy budget is spent in the rural regions of the developing countries, and only 3 per cent goes into renewable energy projects.[25]

Albeit World Bank loans comprise only 3 per cent of global investment in energy, World Bank lending nevertheless exerts a strong influence on the lending strategies of other banks. World Bank lending all but flies in the face of the Bank's own internal analyses: even World Bank experts have long since acknowledged that renewable energy is desperately needed by the majority of the rural population of developing countries. World Bank studies also affirm that renewable energy should be promoted not just on the basis of environmental benefit, but because it also represents the best solution for rural communities. Small, scattered communities do not require central generating capacity or extensive distribution networks. Impressive examples of electrification using local autonomous renewable energy plant are not hard to find – the PV 'solar home systems', for instance, which are increasingly finding application in the rural regions of developing countries.[26] Comprehensive feasibility studies have been carried out for widespread introduction of renewable energy, such as the 'Photovoltaics for the World's Villages' study for the EU Commission.[27] Yet as things stand, no project based on these studies will even begin to do justice to the full scope of needs and opportunities, because this would necessitate a complete volte-face in the industrial nations' policies on development work. The switch to renewable energy has so far been hindered by the relevant decision-making bodies, which are embedded within the framework of interests of the fossil resource industry – and that includes the governments of the developing world countries themselves. Many take their stubborn opposition to this development as a matter of course, being so ideologically bound up in the global energy supply chain that their most obvious opportunity is furthest from their minds.

Perhaps the clearest example of this is the project planned by the South African Development Community (SADC) for high-tension cabling and associated power stations. 1996 saw the signing of a common energy protocol by Angola, Botswana, Lesotho, Malawi, Mozambique, Namibia, Swaziland, Tanzania, Zambia and Zimbabwe.[28] The aim is to construct an energy grid stretching from the equator down to the Cape, which would make it the longest power cable in the world. This grid would be fed from large hydroelectric power stations, some of which have yet to be built, from the coal and nuclear power stations in the Republic of South Africa and from a number of gas-fired power stations.[29] This 'power pool', which would effectively be run by the South African power giant Eskom, is regarded as exemplary, but in reality it is a monstrosity of organized foolishness and cultural devastation. It will cost too much to supply electricity to the villages where three quarters of the region's population live. The high-tension wires will therefore act as a magnet, hoovering up economic activity. The resulting mass emigration from rural regions will leave the elderly isolated in their villages, destroying family structures, and the emigrants' new homes under the corrugated tin roofs of the slums will be a breeding ground for prostitution, dilapidation and violence. The 'modern' concept of bringing people to fossil energy systems, rather than producing energy where they live and can work with – rather than against – nature, merely perpetuates the errors of the past.

Resource dependency despite resource wealth

The countries of the southern hemisphere enjoy the greatest resource wealth, in fossil fuels and minerals as well as in solar

or biological resources. Nevertheless, they remain caught in
the trap of the globally centralized fossil resource industry
because, for better or worse, they have become or been rendered
dependent on fossil supply chains. This can be most clearly
seen in respect of the growing proportion of export income
that developing country economies are forced to spend on
importing fossil energy. The World Bank's World Development
Report gives the following figures for the period 1960–1985.

Table 4.1 *Energy imports as a percentage of export revenue, selected
developing countries*

	1960	1965	1976	1985
Ethiopia	11	8	27	43
Brazil	11	13	28	37
India	11	8	26	30
Kenya	18	No data	54	No data
Madagascar	9	8	22	34
Mali	13	16	25	55
Morocco	9	5	23	50
Pakistan	No data	7	No data	52
The Philippines	No data	12	No data	44
Sierra Leone	11	11	10	63
Sri Lanka	12	11	28	33
Sudan	8	5	26	51
Syria	16	13	16	76
Thailand	12	11	28	33

Source: World Bank: World Development Reports

The table shows the rapid growth in the cost of energy imports
from the mid-1960s on. The two oil crises between 1973 and
1982 cannot have been the sole cause.

The picture would be clearer if more recent World
Development Reports had also given the figures for the period
after 1985. All the signs are that energy imports are consuming an ever-greater proportion of export revenues. The primary
causes are the growing demand for fuel for increasing numbers
of motor vehicles and to supply the additional air traffic generated by tourism. These statistics do not include the proportion
of export revenue that must be spent on imports of motor
vehicles and power plant capable of burning the imported fuel,
nor do they include imports of fertilizers.

Value creation ultimately rests on energy input. If the price of energy skyrockets, then the dramatic conclusion must be that developing countries simply do not have any growth potential, while they remain dependent on imports of primary energy. The import costs eat up the returns that should result from energy use, often before those returns are even realized. The rapidly rising curve from the 1960s onwards suggests that most developing countries now probably spend over 50 per cent of export revenue on imported energy, and in some cases the curve may have already reached or even exceeded 100 per cent.

Energy imports are less of a strain on the economies of the industrialized countries, as trade in energy forms a much smaller part of their economies. This reflects the relatively privileged position that these countries occupy within the global energy system. In 1985, Japan spent 32 per cent of its export revenues on energy imports, Italy 30 per cent, the Netherlands 21 per cent, Sweden and Austria each 18 per cent, Germany 17 per cent, the UK 14 per cent and Switzerland 11 per cent. Among the EU Member States, only Spain and Greece spend a higher proportion of export revenues on energy imports, at 45 per cent and 66 per cent respectively. These countries thus have to contend with disproportionately high energy costs.

Economic development based on fossil fuels is a lost cause, especially for the developing world. The picture becomes even grimmer if we consider which sectors consume the bulk of the imported energy. No statistics have been gathered on this (or at least, none that my lengthy researches were able to locate), presumably because economists are still unaware of the problem. Developing countries expend a considerable proportion of their imported energy on smelting and transporting mineral ore from their mines. These minerals make up more than 50 per cent of exports in many cases, and in some more than 90 per cent (New Caledonia 99 per cent, Zambia 92 per cent, Namibia 77 per cent, Guinea 70 per cent, Togo 66 per cent, Zaire 60 per cent, Morocco 52 per cent).[30] It would be interesting to work out what proportion of the hard currency income from these mineral exports is spent on energy for extraction, processing and transport alone. The value of these minerals to the exporting countries might be seen to be more questionable than it already is – quite apart from the fact that it sends the wrong political and economic signals by promoting a centralized energy system.

In his work *Africa Undermined*, Greg Lanning writes: 'And so it was that "Botswana's" new copper-nickel mine came to be financed by a South African mining group, using a Finnish smelting process and an American refinery in Louisiana. The mine output was sold to guaranteed buyers in West Germany. Perhaps it is pertinent to ask just what contribution this operation will make to Botswana's economic development?'[31] In all likelihood, it is only the employees in the resource industry, along with corrupt politicians and government officials, who profit from the resources business. And it is even more likely that the balance sheet for the fossil energy system, including imported components for power stations and imported motor vehicles, would show an overall loss to the economies of most developing countries. As the global energy crisis deepens and prices rise, these countries are at risk of being economically strangled by the fossil energy supply chain.

5

The mythology of fossil energy

MYTHS, AS DEFINED by the French philosopher Roland Barthes (1915–1980), throw a veil of apparent immutability over the artefacts of human activity. They transform 'anti-nature into psycho-nature', creating an illusion of objectivity that obscures any bias, caprice or vested interest. Once mythologized, a given state of affairs is deemed to admit of no alternative. When confronted with real options, the myth gives rise to 'an artificial delay in which it makes itself at home'. It turns itself into a 'talking corpse'. Myths do not necessarily deny the existence of problems; they 'simply cleanse them and absolve them of guilt'. Alternatives become no more than a Punch and Judy show, so that few dare to defend them, the aim being 'to render the world immutable'. The role of the myth is to delimit the sphere of activity within which people 'are permitted to suffer without changing the world... No more choices need be made, it must simply be endured.'[1]

This is a fitting description of the mythology of the global energy industry. Although they can see the growing dangers, the representatives, political patrons and apologists of the energy industry and its numerous mercenaries and disciples in science and the media remain wedded to fossil energy sources and the specific production and supply infrastructure they require. We do not have the technology for a comprehensive renewable energy supply, they argue, and we cannot afford to switch. The need for fossil energy is a practical constraint that society must respect, for better or worse; whereas proposals for a swift and immediate reorientation towards renewable energy are denounced as irresponsible.

It is the 'economics of energy' that says we must accept the disruption of ecosystems, together with all the other risks associated with fossil energy. The storm clouds may be gathering, the vultures circling over the bodies of the unfortunate, yet the caravan moves on into the expanding desert, while the energy industry, chained to its fossilized structures, continues to protest its indispensability. Rather than putting itself at the service of economic and social activity, it is mythologizing itself as the holy grail of economics.

The theory and practice of economics has always been a happy hunting ground for spurious and contradictory arguments. The more dogmatic the theory, the more serious the absurdities and policy errors that result. The nadir is reached when dogma masquerades as mathematically exact science. As long as the errors that result do not violate the laws of nature in any irreversible way, then they can be corrected at greater or lesser cost to society. A dogma which intimates that we cannot and must not deviate from a course that is heading for the rocks at an ever-increasing speed is, in effect, 'anti-economic'.

Only by recognizing that the fossil energy industry's analyses are inapplicable to renewable energy can we escape from our intellectual imprisonment. 'Energy economics' was developed as an analytical tool for the highly centralized atomic and fossil fuel energy complex. In essence, it is the party-political economics of the energy industry. Energy is all that counts in this analysis, and energy sources are evaluated solely on their capacity and performance, without any consideration of the crucial differences between fossil and solar primary energy, or the economic structures of power generation, supply and use and their differing environmental impact. This energy-economic analysis, however, is blind to the potential for developing and shaping the economy that alternative solutions offer. It puts the specific constraints of fossil energy before the interests of the macroeconomy and of society as a whole.

The following sections offer a critical analysis of the theory and practice of energy economics, and go on to develop a comprehensive framework for analysing the economics of energy that will do justice to the potential for alternative energy sources in the face of macro-ecological challenges. Important new economic opportunities will also be highlighted.

Figures of fancy: the inadequacy of conventional energy statistics

Energy supply and demand is tracked by a welter of official statistics. Yet the picture they paint of the role of renewable energy within our current energy system is misleading at best. The sampling and data-collection methods employed were designed with commercial suppliers of conventional energy in mind, which means that they are blind to much of the potential of renewable energy. Conventional statistics do not reflect reality.

The only types of energy that find their way into the statistics are those for which the industry keeps accounts: extraction and imports of primary energy (ie, coal, natural gas or crude oil); sales of fuel; and the generation and supply of electricity. Different collection methods employed in different countries also give very different pictures. Some countries calculate their energy statistics on the basis of inputs of primary energy into refineries or power stations; others use sales figures for gas, petrol, diesel or electricity, which often makes international comparisons dubious. A country which, for instance, measures the proportion of electricity supplied from nuclear energy by the amount of steam produced from the reactors may record significantly higher figures than a country which measures only the actual current generated by nuclear plants. Another reason why energy statistics are of limited use is that figures, such as those produced by Eurostat (the EU Commission statistical office), only list the ultimate output of electricity, omitting energy losses during the generation process. This omission can lead to erroneous conclusions, especially in the case of environmental evaluations, where it is the quantity of environmentally damaging primary energy consumed that is crucial. Electricity from combined heat and power (CHP) plants with an efficiency of 70 per cent or more cannot be directly compared with current produced by a plant that is only 40 per cent efficient.

Electricity generation from renewable sources also involves losses. However, as wastage of water power, wind or sunlight has no negative consequences for the environment, a complete analysis of the two primary considerations of environmental impact and resource consumption would have to include the real quantitative value of substituting renewable energy for nuclear or fossil fuels – ie, the data would have to show how much primary energy would be saved by using electricity from renewable sources, and not just quote amp-for-amp figures. Measured in terms of primary energy, the real contribution of renewable energy to real energy consumption is clearly higher than shown by the energy statistics. Calculations performed by Wolfgang Palz from the EU Commission, for example, suggest that the proportion of energy generated from renewable sources in the EU exceeds the Eurostat figure of 3.7 per cent or 45.4 million TOE for 1991 by more than a third. The substitution value is 68 million TOE.[2] Even this figure is too low if the energy losses prior to and following combustion of fossil fuels in power stations are also taken into account. According to the Stockholm Environment Institute, the additional losses for crude oil are 2 per cent during extraction and transport to the refinery, and a further 8 per cent in the refinery itself, as measured by comparison with the quantity originally extracted. In the case of gas, pre-combustion losses are 10 per cent, and 7 per cent in the case of coal. If these fuels are subsequently used to produce electricity, this results in additional losses of 8 per cent.[3] This means that across the whole supply chain, the losses go beyond the 60 per cent loss within the power station, amounting to 69 per cent in the case of oil and gas, and 67 per cent in the case of coal. If the oil is used to fuel motor vehicles, then losses rise to 90 per cent, according to energy flow calculations by the Luxembourg firm MDI (Motor Development International). None of this is reflected in the official energy statistics, which are effectively an instrument for pulling the wool over the eyes of decision-makers and the public.

The simple comparison of fossil fuels with biomass commonly employed in compiling statistics is also thoroughly misleading. It is insufficient simply to compare the respective numbers of tonnes consumed. The figures for biomass usually relate to dry mass, excluding the weight of the water in the material when it was first harvested. The figures for fossil fuels, by comparison, relate to the actual quantities extracted, which implicitly assumes that processing for use as fuel or industrial raw material is loss-free. It also implicitly denies that plant material can be productively exploited without first extracting the water content.

But the statistical deficiencies in the evaluation of renewable energy go far beyond the examples so far discussed. Many of the ways in which humanity benefits from the sun's energy are not accounted for at all, partly because the statistical data would be too costly too collect, but also because the conventional understanding of energy is too narrow.[4] For example, there is an average quantity of energy that people need to be comfortably warm. In warmer climes there is considerably less need for heating, because the sun provides the necessary heat. This reduced need for heat energy enjoyed by a large part of the world's population ought to be included in world energy statistics as part of the contribution made by renewable energy. The quantities of energy involved are very significant, and should not be lightly disregarded.

Solar energy input in terms of the seasonal variation in conventional energy demand between summer and winter is likewise taken for granted, and not measured. Demand for oil and gas for heating systems in central Europe is, for example, overwhelmingly concentrated in one half of the year, yet it still makes up – along with the fossil energy used to provide hot water all year round – 40 per cent of total statistically observable energy consumption. In the months when no fossil heating energy is required, the sun meets all heating needs. Nevertheless, despite, or perhaps because of, its ubiquity, this solar heat input is omitted from the statistics, as if it were of no practical importance.

Heating needs met from solar collectors or from wood-burning stoves, houses positioned for maximum solar gain, conservatories, transparent insulation, double glazing, exploitation of the heat input from the human bodies living or working in the building, heat exchangers, exploitation of ground heat –

none of these solar heating gains find their way into the energy statistics.

Further examples: people need artificial light from sunset to sunrise; during the daytime, the sun meets lighting needs. The need for artificial light is lower when the days are long than when the days are short. The difference in electricity demand between these two times of year is an indication of the proportion of lighting needs that is met by the sun.

These energy inputs are ignored because they are taken for granted, yet they are of great practical importance. There is considerable scope for energy conservation through using town planning to maximize building solar gain, and through architectural features and additional 'daylighting' technology that allow a maximum amount of daylight into the building. Nevertheless, energy statistics take no account of energy-conscious planning and design. The same can be said for the replacement of cooling systems powered by diesel motors or grid electricity with natural cooling and flexible shading.

The inadequacy of energy statistics also extends to the figures for electricity consumption: what does not flow through the grid does not get counted. Not a single form of autonomous energy generation is recognized in the energy statistics! Yet the range of autonomous systems extends from wristwatches to pocket calculators, from water pumps to autonomous houses with no grid connection, from solar lamps to street signs lit using PV, from solar-powered battery chargers to the solar home systems in developing country villages and small-scale wind turbines. This list could be extended indefinitely, both for heating and for electricity. It also includes sun-dried crops, irrigation windmills, biological fertilizers, cycling, solar-powered boats and many more examples of how fossil fuel consumption can be avoided or replaced. Chapter 6 deals with autonomous systems, which offer an extremely large – indeed presumably the greatest – technological opportunity for replacing fossil fuel or nuclear energy. Ignorance of the economic importance of these energy inputs extends even to figures of speech, as used unconsciously even by advocates of solar energy. For example, one term in common use is 'zero-energy house', although taken literally this is a physical impossibility. What is meant is a house that derives its energy solely from the sun and which thus has no need for *additional* fossil fuel energy. 'Zero-emission house' or 'autonomous house' would be more accurate. The term 'zero-energy house' can only perpetuate the misconceptions surrounding solar energy.

These blind spots mean that energy statistics oriented towards commercial piped energy can provide only a fragmentary and thus wholly inadequate understanding of energy. They obscure the fact that, notwithstanding massive consumption of fossil fuel and nuclear energy, the sun is still humankind's largest single energy source. They divert attention away from the innumerable opportunities for replacing nuclear and fossil fuel energy with solar energy discussed in Chapter 6. With current data-collection practices, theoretically it would be possible to replace more than half of all fossil energy consumption with solar technology without significantly increasing the statistically observable proportion of energy demand met from renewable sources. Renewable energy successes are thus systematically excluded from the statistics. Energy statistics are in any case produced by the atomic and fossil energy industry, and are presented only in abbreviated form. The figures do not record energy use; they record nuclear energy, crude oil, gas and coal use. The bounteous flow of energy from the sun, and the opportunities it represents, is thus swept under the carpet with mathematical precision.

The inadequacy of energy forecasts

The usefulness of the various forecasts for the expected growth rates of individual energy sources is also at best limited to the fossil energy industry itself. Current forecasts are effectively useless for predicting the likely future role of renewable energy because they obscure the crucial structural differences between conventional and renewable energy. Figures published regularly by the Association of German Electricity Producers (VDEW) on the expected future composition of the energy mix amply demonstrate the arbitrary nature of renewable energy forecasts. The VDEW member companies are questioned about their investment plans, and the answers are a product of the compa-

nies' expectations of future demand, estimated lifetime of existing capacity, long-term supply contracts with other energy
companies and plans for future investment, including investment in renewable energy. Forecasts are derived from the
responses received, and these also form the basis for estimates
by academic institutes.

Forecasts relating to the energy industry thus constitute a
self-referential system. Where forecasts differ, this is usually at
most the result of different assumptions concerning general
economic trends or price movements on the markets for
primary energy. If the assumptions made fit the trend for the
following ten or twenty years, then the forecasts may well be
correct – but only as long as renewable energy is kept out of
the picture, or only considered to the extent that the dominant
energy companies are prepared to invest in it.

The energy industry is also tacitly assuming that what it
regards as its hereditary monopoly on energy supply can be
extended to cover renewable sources as well, and this is reflected
in the industry forecasts. The many restrictive forecasts predicting only a limited role for renewable energy in the short and
medium term are thus effectively devoid of informational value.
At most, all such forecasts can tell us is that the forecasters and
their sponsors lack the interest, ambition or sufficient imagination to challenge the conventions of fossil energy. Frequent
statements to the effect that solar energy 'will' only make a small
contribution to energy supplies in the foreseeable future, and
that it 'cannot' replace fossil fuels, are no more than pseudosci-
entific excuses for the structural conservatism of the energy
industry. Even public forecasting institutes such as the IEA
connive in this, when, for example, it forecasts the proportion of
world energy supplies coming from renewable sources in 2020
to be 3.1 per cent, as compared with 3 per cent in 1995 – including large-scale hydropower, but neglecting the non-commercial
use of biomass that plays a large role in the developing world.[5]
True, renewable energy use is expected to grow by 1.5 per cent a
year in absolute terms, but as overall energy demand is also rising,
the net effect is only a slight increase in relative terms.

Two of the most important global energy companies, the oil
giants BP and Shell, have recently broken the conventional

forecasting mould with much-cited studies predicting considerably larger proportionate shares for renewable energy. The study by Shell talks of a 50 per cent share for renewable energy by 2060.[6] Renewable energy enthusiasts have made much of this study, which does go some way towards overcoming the prejudices against renewable energy. 'Well, if *Shell* says so...' Yet as instrumental as this study has been in breaking the ice for renewable energy, and as right as it is to acknowledge that renewable energy can make an at least equal contribution to global energy supplies, even this forecast is too much a reflection of the ambitions of the company that published it for claims of scientific rigour to have any merit. Besides the greater role for renewable energy, the Shell study is also forecasting a doubling in the global energy demand, of which half is to be met from fossil sources. Shell is therefore broadly expecting oil and gas consumption to remain at the current level through to 2060. As Chapter 3 explained, however, consumption at this level is unrealistic because at current levels of demand, the oil will be gone by 2060. The Shell study is careful to avoid the subject of replacements for fossil fuels. The reasons are probably tactical – the company has one eye on its fossil energy business, and the other on its shareholders. Corporate strategy seems to be to open the door for a new business model based on renewable energy, but without delegitimizing the existing fossil energy business. In recent times, through their respective newly founded subsidiaries Shell Solar and BP Solar, Shell and BP have become the biggest single investors in renewable energy. BP, which got off the mark faster than Shell, has announced its intention to invest $20 million annually to 2010 in renewable energy; Shell promises $500 million in five years. But at the same time, according to data from German Watch, BP invested $4 billion in fossil energy in 1997 alone, and Shell $7.5 billion. There are sound business reasons for this: the finance for the investments in renewable energy must come from the fossil energy business. In any case, being tied into the web of fossil energy supply chains leaves both these global corporations with little choice other than to stand by their existing fossil energy businesses.

It is no coincidence that the oil companies are the first of the global players to break the hermetic seal on renewable energy: crude oil will be the first fossil fuel to run out. For decades, it was these very same oil companies which sought to block political initiatives promoting renewable energy – such as in the infamous Global Climate Coalition, which attended climate change conferences to lobby against political decisions to limit CO_2 emissions. BP and Shell have since left the coalition.

Whether optimistic or pessimistic about renewable energy, energy forecasts almost always lack scientific credibility. Renewable energy simply does not fit into a forecasting methodology designed for fossil fuel structures. Investment in renewable energy takes the form of innumerable small projects, many of which fall outside the supply chains from which the energy statistics are drawn. How fast and how widely the potential of renewable energy is realized depends on the motivation of many millions of people, on the level of energy awareness in the public consciousness and, above all, on the extent to which political action widens the scope for exploiting renewable sources: by eliminating legal barriers to renewable energy through genuine energy taxation, by energy market regulation and through pump-priming programmes. At the moment, the wind of change blows at different strengths in different cultural and legislative environments. Renewable energy is correspondingly gaining ground in Europe at different speeds in different Member States. In mid-1999, Germany had installed around 3500 MW of wind turbine capacity, and Denmark 1560 MW, whereas France had installed only 19 MW and Ireland 73 MW – although the Atlantic coastline endows France and Ireland with many more suitable sites for windfarms than Denmark or Germany. The difference is not down to geography, but to the favourable climate for wind turbine operators in Denmark and Germany provided by the 'electricity feed-in laws' (*Stromeinspeisungsgesetz*) that guarantee grid access and minimum prices for renewable energy. For reasons of politics and culture, there is also clearly more enthusiasm for environmentally motivated private investment in these countries.

Italy, for example, enacted an electricity feed-in law at the same time as Germany, with similarly favourable tariffs for renewable energy, but the act had practically no effect in terms of investment by independent operators. The hegemony of the Italian state electricity company ENEL remains unbroken. Greece has a population only one sixth of Italy's, lower gross domestic product (GDP) and lower average incomes. Nevertheless, Greece has 80 times as many solar collectors installed as Italy. Even in Germany, Denmark and Austria, there are more solar collectors than in the largest and richest country on the Mediterranean. All this goes to show that costs are far from being the decisive factor in determining when, where and how much solar plant is installed.

The prerequisites for investment in fossil fuels, which are tied to supply chains, and in renewable energy are fundamentally incomparable, and it is this which determines the validity of forecasts.

- Fossil fuels are largely a mature technology. Large power plants have long lead times; no further large increases in efficiency are to be expected; power plant construction is a mature and established industry; the infrastructure for transport and distribution is largely already in place in the industrialized countries; marketing structures are well established and mostly under monopoly control.
- In the case of renewable energy, there are many technical advances still to be made; the existing technology is still immature and offers considerable scope for development; large increases in efficiency can be expected; plant manufacture is a new and growing industry; and the history of other technologies shows that radical optimization of production techniques and steeply falling costs through mass production of standardized products can be expected. Marketing structures are only just beginning to emerge.

It is equally difficult to foresee how significant a role will be played by different usage strategies. Besides the considerations detailed above, it depends on the specific geographical conditions and on which technology for exploiting renewable resources first makes an industrial impact and leads to reduced costs. What can be said is that there will in principle be considerable differences from region to region and continent to continent in the way that energy needs are met, and thus that energy systems as a whole will become much more diverse.[7]

Calculations show that the entire global energy demand can be met from renewable sources.[8] Everything else depends on spurs and initiatives from the technological, business and political spheres. Renewable energy does not just require different patterns of use; it also need new sponsors and a new investment culture.

It is not scientifically possible to predict how fast nuclear and fossil fuel energy can be replaced by renewable energy, simply because it is not possible to anticipate future developments in solar technology and the behaviour of potentially billions of energy consumers. The share of renewable energy in world energy supplies by the middle of the 21st century could be less than forecast by the Shell study; equally, it could be far higher, even to the extent of completely replacing nuclear and fossil fuel energy. That is the goal that must be pursued, and it is by no means unrealistic. The faster the shift occurs, the longer limited fossil fuel reserves will last. At the same time, fossil fuel costs will be rising as the returns to scale (the cost of energy supplies in proportion to the costs of maintaining the conventional energy supply infrastructure) begin to fall. All forecasts that go beyond the next 20 years are in any case obviated by the goal of rendering conventional energy ever more expensive by restricting dwindling supplies, reduced economies of scale and proper energy taxation regimes, while lowering the cost of renewable energy through a range of multipliers and continually improving returns to scale. If renewable energy seriously begins to come on-stream and roll-back attempts cease, no long-term prediction will worth the paper it is written on.

The persistence of objections and counters to optimistic visions for the future – such as dispensing entirely with nuclear power and fossil fuels – has one key cause: to acknowledge the plausibility of such visions would be to explode the mythology of the energy industry. There could no longer be any justification or acceptance for continued large-scale investment in fossil fuel and nuclear plant. There would be equally little acceptance for pouring further billions of public money into the development of nuclear fusion solely on the basis that future energy needs cannot be met from renewable sources alone.[9]

The profligate subsidies for conventional energy systems

Lured by sirens singing of lower energy prices, civilization is allowing itself to be drawn ever faster towards ruin. The centralized energy industry is using dumping prices, particularly on the open energy markets, to prove its contention that it is the most cost-effective system for energy supply possible, and thus an indispensable element of the aggregate economy. But such arguments can fall on fertile ground only because the energy discussion is being conducted in fuzzy terms. A case in point is the way the nuclear/fossil fuel energy complex regularly presents its current competitive advantage as a fundamental economic advantage. To reach this conclusion, everything that occurs before and after the direct generation costs is simply disregarded, including direct and indirect, current and past subsidies, the drain on the economy caused by imports of primary energy, the cost of permanent destruction of resources, and environmental costs. By repudiating the crucial distinction between business and macroeconomic accounting, the energy industry is claiming to be the sole competent authority for energy issues, while at the same time denying any liability for the economic, ecological and social consequences of its actions.

Energy subsidies: the economic bankruptcy of conventional energy systems

The multiplicity of direct and indirect state subsidies is proof positive that the economic viability of nuclear and fossil fuel energy is founded on deception. It is not just the immense start-up grants given to the nuclear industry in times past, state aid on a scale never granted to renewable energy. The largesse lavished by the public purse on the nuclear and fossil fuel industries goes far beyond the funding of research and development. It extends from measures to support the market and underwrite investment in infrastructure through to subsidized energy for large companies by means of numerous tax and insurance privileges and the provision of free civil and military security services, to say nothing of picking up the tab for the subsequent cost of damage to environment, human health and the climate. It is scarcely possible to calculate the extent of these knock-on costs and subsidies; they are probably comparable to the annual worldwide defence expenditure of $850 billion. The state is waging war on the world's ecosystems and cycles.

A relatively small proportion of the total expenditure, but still disproportionately high, consists of the subsidies for research and development in the nuclear and fossil fuel industries. My book *A Solar Manifesto* lists subsidies for all OECD countries between 1984 and 1995 of $9.27 billion for renewable energy, $17.48 billion for fossil fuels, $56.43 billion for nuclear fission and $14.64 billion for nuclear fusion.[10] These figures do not include expenditure on nuclear weapons research by countries with nuclear capability, from which civilian nuclear power also benefits (and vice versa). A study published in 1997 by Greenpeace summarized direct state subsidies – ie, for research, development and market support – for the Member States of the EU. In 1995 the figures were $9.68 billion for fossil fuels, $4.1 billion for nuclear power, but only $1.24 billion for renewable energy.[11] The Alliance to Save Energy puts US state subsidies in 1994 at at least $21 billion, of which around 95 per cent was spent on nuclear and fossil fuel energy.[12]

These figures do not include the innumerable hidden subsidies. Most significant is surely the lack of excise duty on civil aviation, which – estimating from average rates of duty on road fuel – may weigh in at well over $100 billion a year, increasing as the aviation industry grows.[13] The lack of duty on fuel for international shipping is of a similar order, although, as with aviation, there are no statistical data on this. The total cost is probably about the same as for air travel. Further examples are the fuel duty exemption in the EU for oil-processing companies, which effectively subsidizes not just the refineries, but also the chemicals industry. The tax system also gives preferential treatment to prospecting for and opening up oil and gas fields, and to subsequent extraction activities. The same goes for uranium.

Among the hidden subsidies are also state co-financing of port facilities and the construction of pipelines and high-tension cables. Steven Gorelick provides many examples from a variety of countries in his book *Small is Beautiful, Big is Subsidized*.[14] Almost all countries with a nuclear industry also subsidize operators of nuclear power plants by means of generous exemptions from public liability in the event of nuclear accident. In 1988, the USA did raise the legal requirement for indemnity cover from $560 million to $7 billion, but still retained the concession that stipulates that the insurance premiums are payable only in the event of an actual incident – ie, retrospectively.[15] And even $7 billion is still extremely low when compared with the final bill for the Chernobyl explosion of $350 billion – a sum that does not take into consideration the suffering of those with terminal radiation poisoning, which of course cannot be expressed in monetary terms. Further hidden subsidies are the policing of atomic installations and nuclear waste transports, and military security for oil extraction facilities, which the US organization Citizen Action puts at $57 billion annually for the USA alone.[16]

The United Nations Development Programme (UNDP) published a report in 1997 entitled *Energy after Rio*, which mentions annual subsidies of $300 billion for conventional energy. This includes subsidized prices for developing-world consumers for whom the world market prices are too high. 1994 figures from the World Bank put the cost of price supports in the developing world alone at $90 billion. The lion's share of these subsidies, as detailed in Chapter 4, go to benefit the urban population.[17] The UNDP studies also do not consider the full range of hidden subsidies listed above.

Economic analyses of conventional energy systems in the electricity sector also fail to take account of regional monopolies. As long as these were – or, new energy market legislation notwithstanding, remain – in place, there was no need to cost investments accurately. Monopolies can pass all their investment costs on to the consumer at no additional risk, which is why the regional (ex-)monopolists have a large pool of power stations at their disposal for which depreciation charges have already been paid. The electricity companies are therefore de facto subsidized by their customers, a position they can use to beat off new entrants who, because of the high initial investment costs, cannot match sinking electricity prices. Even within a market that is officially free and open, the large electricity companies can still maintain and even extend their position by using dumping prices to fend off new entrants to the market. The free market for electricity is dominated by oligopolistic competition between large firms. As long as the traditional excess capacity of these firms allows them to forgo new investment, prices can continue to fall. But as soon as this phase is over and the process of monopolization is further advanced, the need for additional investment alone will bring large price increases. The aim is to head off competition from new entrants and municipal power companies before this point is reached. The scale of consumer subsidy of electricity companies was revealed during the discussion on 'stranded investments' following the opening of the electricity market. The term 'stranded investments' refers to investments undertaken regardless of actual demand. The total value of such stranded investments was estimated at $50 billion in the USA alone.[18] The electricity companies' attempts to disguise this overinvestment by seeking to persuade governments to underwrite sales – for example, in the case of electricity from lignite-fired power stations in the former East Germany – expose the bankruptcy of their business model. Such legal protection is accepted as a matter of course in the established electricity industry, but not for renewable energy or for municipal power companies.

One might object that the tide of subsidy for nuclear and fossil fuel energy does not wash equally highly in all countries, and that the existence of subsidies alone is insufficient to disprove the argument that fossil energy is fundamentally more cost-effective than renewable energy. But as the analysis of conventional energy supply chains in Part I indicates, only the sheer scale of production and subsidy makes it possible to absorb and disguise the innumerable cost-centres of fossil fuel and nuclear energy. The larger the quantity of subsidized energy supplied locally, the lower the global price, wherever the energy is consumed. The lower the price, the greater the flow of energy. Wherever energy production and supply is subsidized locally, the subsidies paid reduce the costs of fossil fuels globally. If all subsidies were to be withdrawn, the resulting price hike would perhaps make state support for renewable energy unnecessary. The unilateral removal of subsidies by one country alone, however, would not be enough to overcome the lead enjoyed by conventional energy. Only the lack of accounting transparency in global commodity supply chains allows the myth of the lower cost of fossil fuel energy to be preserved.

The feigned productivity of nuclear and fossil energy

Businesses seek productivity growth for a number of reasons, of which price competition, whereby high costs may cause a business to be out-competed by more productive rivals, is only one. Other motivations include the quest for increased profitability and ways to make tasks easier and simpler to perform, greater user-friendliness for a more competitive product, reduced environmental impact and time savings. Which of these motives dominates will depend on the particular conditions obtaining. While energy was still expensive, efforts to increase productivity concentrated on energy efficiency. Since human labour has become expensive, the focus has been on automation.

Disproportionately low conventional energy prices are most frequently cited as the reason for highly suboptimal energy productivity, but there are other reasons as well. There is the purely ideologically motivated fixation of the business world on current prices, as befits the neoliberal mentality of microeconomic calculation. Furthermore, the concentration of the search for productivity gains on technological solutions is extremely short-sighted, because it fails to address wider questions of the relationship between energy and society.

The fixation with current prices

In this supposedly modern age, the sole criterion for evaluating the viability of a particular source of energy is almost invariably its market price. Market prices are assumed to reflect current costs, and indeed the two terms are used more or less interchangeably. That, however, is an anachronistic analysis of economic viability. Cost and price are by no means identical. The whole history of economic development since the Industrial Revolution shows that, time after time, increases in productivity with the aid of judicious use of energy and efficient generation technologies have made it possible to reduce energy costs despite stable or even rising prices. Equally, the extremely low energy duty in the USA, among other things, has given the country the lowest energy prices in the OECD, but by no means does that automatically translate into lower energy costs for households or industry. Low prices encourage markedly higher energy consumption and provide no incentive for investment in energy efficiency. No wonder that US citizens consume two to three times as much fuel and electricity per head, effectively negating all price advantages.[19]

Equating – or confusing – prices with costs is an argument from the pre-technological age, and an expression of structural conservatism. Nevertheless, this is the argument that dominates the energy debate. Any mooted increase in energy taxation is subjected to a barrage of criticism, on the grounds that the consequent energy price rises must necessarily mean equivalently higher costs, which in turn would endanger the economy's international competitiveness. The immediate response in those countries which have instituted environmentally motivated duties on energy, be it Germany, the Netherlands or Denmark, is to grant exceptions for energy-intensive industries – despite the fact that there is much to suggest that it is precisely in industries with above-average energy demands that the greatest scope for efficiency gains lies. This obstructive attitude towards energy price rises permeates even international comparisons of energy costs, which in fact do no more than simply compare prices. Such comparisons reveal little information about the economies compared. In order to account for productivity differences, it would be more germane to compare the proportion of costs attributable to energy in private households and comparable manufacturing and service industries. The absence of such statistics results in systematic errors of judgement both in energy policy and within industry, errors which bind society to conventional energy and make the discussion of vital fundamental change taboo.

The low productivity of centralized energy supplies

The analysis of global energy supply chains presented in Chapter I will have made clear that while nuclear and fossil fuel energy supplies can be managed more effectively, they can never be truly productive. Conventional efficiency calculations, of course, make no mention of this. In the case of electricity supplies, only the input/output efficiency of the generation process is considered. No account is taken of energy losses over the whole supply chain, or of losses during the construction of drilling rigs, ports, pipelines and power stations.

Yet the structural productivity gap of centralized energy systems goes deeper still. For example, the nominal generative efficiency of a large power station applies only if current is actually being produced from the fuel consumed, which is not always the case. Power stations have to cope with fluctuations in demand which can never be accurately predicted, so there must always be steam on tap to drive the turbines – which means that fuel must be burnt even when demand is low. If demand falls, the steam must be vented. Depending on the actual load on the power station, further energy losses are thus inevitable. Steam turbine power stations can achieve their optimum efficiency only if demand remains constant, which is why base-load electricity is the cheapest. Underutilized capacity and superfluous fuel consumption are corollaries of large-scale power plants.

Local micropower plant does not suffer from these problems. If small motor-driven generators, which can be switched on and off in seconds, are used in pace of large steam turbine plants, then there is no need to maintain heads of steam behind turbines, and no need for reserve capacity. Small power units, typical for most forms of renewable energy, make for a modular system that can be tailored to meet market demand à la carte. There is much less risk of misplaced investment. Every module is independent, and short lead times make it possible to react quickly to increased demand. Investment returns kick in immediately.

Looking for environmental efficiency in isolation

Environmental considerations are also used to argue against a swift re-prioritization in favour of renewable energy. Environmental gains, it is claimed, can be achieved more quickly by saving energy or using fossil fuels more efficiently than by costly investments in renewable energy. At first glance, this argument seems convincing, and it can no doubt be backed up by calculations in many cases. Nevertheless, to apply it generally would be at least questionable, if not outright absurd. The environmental efficiency of investment in renewable energy is in many cases comparable with investment in energy efficiency. That is the case, for example, with the 'passive' use of solar energy in buildings. Even the German environment ministry[20] has frequently brought up energy efficiency as an argument against the use of vegetable oil as fuel, preferring to aim for the introduction of fuel-efficient vehicles (the so-called 'three-litre car'). This overlooks the fact that motors which run on vegetable oil can be just as fuel-efficient as diesel or petrol engines. Fuel efficiency as an argument against vegetable-oil fuel simply prioritizes fossil fuels over renewable energy, which is environmental nonsense. Even for PV, still the most expensive solar technology, there are cases where the environmental cost-effectiveness argument no longer applies. Where they make the construction of distribution grids and cabling unnecessary, PV installations are already often more cost-effective than all forms of conventional energy. Arguing against the exploitation of renewable energy on energy efficiency grounds is irresponsible environmental and development policy. Whether it is more appropriate to invest in more efficient use of fossil fuel energy or in renewable energy, or in both, will depend on the specifics of the case in question.

Even where more efficient use of fossil fuels will bring greater immediate environmental returns, one must then ask whether this is still the case once the entire life-cycle of the technology has been taken into account. It is not enough simply to compare the current cost of investing in fossil fuel or solar plant; the running cost of fuel for the more efficient fossil fuel plant must be compared with the zero input costs of a solar installation. In any case, there is a limit to how efficient fossil fuel plant can be. As a rule, the marginal cost of efficiency improvements increases with each additional saving, whereas the price of renewable energy technology falls with increasing market penetration. The decisive factor in the economic analysis is the direction of the cost trend. This must be taken into account when extrapolating into the future.

In a society composed of independent economic agents, it also makes no sense to assume that the cost–benefit analysis in terms of safeguarding climate and environment will always favour investment in energy efficiency. What are farmers wishing to reclaim their agricultural waste using a biogas plant, or to erect a windfarm on their fields, to make of such a blanket generalization? Or householders who, having exhausted all the energy efficiency options open to them, now want to generate their own electricity from PV? Are they supposed to forgo investing in a project with which they identify and which is within their power to realize, in favour of some anonymous investment in more efficient use of fossil fuels? If the argument that greater energy efficiency brings greater environmental benefits were to be followed to its logical conclusion, there would need to be a central bureau for all energy investment, whose responsibility would be to allocate available capital to the most effective investment. That may sound like a crude caricature, but the conclusion is implicit in the efficiency dogma of some studies. Economic trends require a variety of motivations; reducing all motives to the level of unconditional cost–benefit calculations stifles individual dynamism and promotes conformity.

The received wisdom of fossil fuel economics seeks to trump every renewable energy initiative by asking whether it 'pays'. But how much of human activity would cease were this to be the sole criterion for spending money? From house and flat décor to sunny holidays abroad, from eating out to stylish cars – whether any of these are worthwhile is down to individual taste and priorities. Clean energy is an emotional and ethical need as well as a rational one. Relativizing this need by reference to up-front costs is a mistake that even proponents of environment-friendly energy can be persuaded to make. Energy is not a special case; there is no reason why consumers should treat it other than they would any other commodity or good. Equally, there is no reason why energy supply should remain the preserve of the established energy industry alone.

The fundamental inefficiency of fossil fuels

The sun is the ultimate origin of all known energy sources. Oil, gas and coal are derivatives of biomass produced by the sun over a period of around a billion years. Geological processes such as pressure and the exclusion of air converted this biomass into the form that is extracted and burnt today. However, as only a few millionths of the original biomass were converted into coal, oil and gas, only 0.000011 per cent is available today as a source of energy. By comparison, once biomass harvested today has been dried, its energy content is available in full.

Such considerations are more than purely theoretical. The logic is that of the national accounts, in which an increase in the money supply is equated to growth. Everybody acts as if reserves of fossil fuel and the Earth's capacity to absorb waste and emissions from power generation were unlimited. This is no formula for growth, but rather a twofold loss, of resources on the one hand and environmental quality on the other. Because nature is not an accountant and sends no invoices, the incalculably high cost of consuming fossil fuels is overlooked. Fossil fuels necessarily represent a departure from the variety and multifunctionality of their solar origins. The broad spectrum of solar irradiation, from ultraviolet to infrared, can be put to a variety of different uses, from light for the production of electricity to the use of infrared radiation for heating. By circulating a thin film of water over its sunny side, a solar cell can also be made to serve as a solar collector, raising its efficiency from the current 10–15 per cent to 50 per cent or more. As these 'sunlight harvesters' can take the form of building components for roofs, facades or windows, for fences or balconies, they can be made to serve far more functions than has hitherto been the case. The wide variety of ways in which wholly new solar energy systems can be used gives rise to a completely different set of efficiency calculations.

The scale of what can be achieved can be seen in nature. A single tree absorbs CO_2, produces oxygen, prevents evaporation and serves as a reserve of resources and energy; it can produce food, serve as a wind break or protect against erosion –while also being nice to look at. Only solar resources can achieve such multifunctional efficiency. Nature sets the standard for the technological and economic realization of the potential of solar resources discussed in Part III.

Ideology and the physics of energy

Most physicists to this day regard the achievable potential of renewable energy as insufficient and of little use. They believe that it is impossible to replace all existing fossil fuel and nuclear energy supplies, arguing that fossil or nuclear energy must be available to cover for when the sun does not shine or the wind does not blow. They do not appear to have hit on the simple notion that it is also possible to use renewable energy to cover for interruptions in supply. It is in any case already common practice to take capacity on- and off-line to suit varying levels of demand, even if for different reasons. The claim that the base load cannot be met from renewable sources has also long since been empirically debunked.[21] The question is why these arguments stubbornly continue to circulate, even among physicists and the wider scientific community. Even politicians with only the most rudimentary understanding of physics point to the 'laws of physics' when criticizing supposedly overblown expectations for renewable energy, a tactic designed to lend weak arguments the air of scientific profundity.

Physicists' opinions are shaped not just by physical laws, but also by the received wisdom of the time. Armin Witt introduces his book on suppressed inventions with the ironic observation that 'our physical laws say that the bumblebee cannot fly – but nobody told the bumblebee.'[22] One familiar argument against the achievability of meeting all human energy needs from solar energy sources is their low energy density. The term refers to two concepts: energy content per unit mass on the one hand, and the geographical footprint on the other. One cannot conclude on the basis that one tonne of crude oil or coal contains more energy than one tonne of biomass that it is not possible or not feasible to generate energy from biomass. It simply means that biomass is more costly to transport, and that shipping distances must therefore be kept short. The geographical concentration of large fossil or nuclear energy flows in large-scale power plants is by no means essential in order to meet mass demand. Whether the electricity 'pool', in the form of the potential maintained in the distribution grid, is 'filled' from a few large power plants or numerous small ones is immaterial as far as the electricity consumer is concerned.

The size of the grid is equally unimportant. Whether international or national, regional or local, the only thing that matters is that enough electricity is pumped in to meet current demand. The impression given by comparisons between the footprints of the various generation technologies is thus highly misleading. In Germany, the figures are 0.1 kW/m^2 for PV, 3 kW/m^2 for wind power, 500 kW/m^2 for coal and 650 kW/m^2 for nuclear power. Statistics like this are designed to create the impression that large-scale electricity generation requires large-scale power plants capable of producing large quantities of electricity in a very small space. The figures for coal and nuclear power, however, fail to account for the land requirements of the entire supply chain from primary energy extraction to the power station, electricity distribution and waste disposal. Arguments based on energy density are no more than energy prejudice dressed up as physical fact.

All that relative energy density tells us is which structures are required to support the various generation technologies. High energy density generally means centralized structures; low energy density, decentralized ones.[23] Those who accept the need for high energy density lack the motivation or technical imagination to envisage anything other than large-scale production. Why is it that even intelligent physicists venture out on such thin ice? Why do some many of them actively reinforce the mythology of centralized nuclear and fossil energy supplies? Why do respected physicists and even the venerable German

Physical Society founded by Max Planck endorse the techno-
phobic disparagement of renewable energy?

One example among many is the book *Die Energiefrage* (The
Energy Question) by Klaus Heinloth, a respected physics
professor who sat on the German parliamentary commission
of inquiry dealing with the issue of energy supply. Heinloth
also draws on the work of the International Panel on Climate
Change, a UN organization that provides scientific backing for
the declarations of the Global Climate Convention and which
clearly recognizes the global risks posed by fossil energy
consumption.[24] Heinloth attempted to calculate the 'realizable
potential' for renewable energy, in Germany and across the
world, which he believes to be 'maximally exploitable' – ie,
achievable in the optimum case – by 2050. He concludes that
for central heating and hot water, this is two thirds of future
demand, for motor fuel 10–15 per cent, for electrical energy
20 per cent, and for (high temperature) process heat 'as before
only negligible'. The contribution of renewable energy to global
energy supplies could reach 10 per cent for heating and process
heat, 30 per cent for motor fuel, and 30 to 35 per cent for
electricity, 'in the favourable, optimistic case', 'as long as' hydro
capacity is doubled between 1995 and 2050, 200,000 MW of
solar thermal plant is installed in the tropics, wind capacity is
increased a hundredfold (from around 3000 MW installed
capacity in 1995) and 2000 km^2 of solar panels are installed.

Heinloth does go further in his assumptions than many
other physicists. Yet he provides no credible explanation why
the total production of generative capacity in the form of wind
turbines to 2050 will be no greater that Germany's current
annual car output, or why there should be fewer solar panels
worldwide than there are roofs in Germany alone, or why only
10 per cent of global heating needs can be met from the sun,
despite the fact that the majority of the global population live
in the sun-rich South, and that even in northern Scandinavia
whole towns are meeting 50 per cent of their energy needs from
the sun; or why he estimates that only 20 per cent of electricity demand will be met from renewable sources in Germany,
assuming that demand remains constant to 2050, when that
could be achieved using current technology with no more than

30,000 1.5 MW wind turbines. Ten thousand units were installed between 1990 and 2001 alone, albeit initially still with limited capacity. In an 'energy memorandum' in 1995, the German Physical Society also called for a third of electricity to be supplied from renewable sources by 2030, which in its opinion could only be achieved by importing large quantities of energy from solar thermal plants in North Africa.[25] It remains thoroughly unclear why all these figures were not set higher. There is at least no physical justification for them. Instead, the studies base their pessimism on the supposedly limited capacity of the economy to absorb higher costs. In using such arguments, however, they are straying outside their field of expertise and simply accepting without question statements by the established nuclear and fossil energy industry to the effect that costs will rise towards 2050. Enough has been said about the reliability of such forecasts already. The German Physical Society is basing its supposedly scientific conclusions on arbitrary and unfounded assumptions.

The Italian science historian Federico Di Trocchio shows how often new developments have been rejected even by established scientists. As he sees it, their mistrust derives from 'the very structure of the previous theory'.

> As theories cannot by themselves continue to exist without scientists to support them, in practice, it is precisely the adherents of existing theories who invoke pseudo-necessities to criticize the work of other scientists, especially colleagues looking for new answers. This is the psychological and cognitive source of the stubborn refusal to let old theories go, even when those theories have long since sunk to the level of prejudice.[26]

The old energy theories have also been reinforced by the perception that renewable energy represents a retrograde step in scientific and technological terms. Progress has always been and continues to be looked for in ever grander and more complex technologies. In the case of atomic physics at least, these technologies doubtless presented greater challenges to the physical sciences. According to this line of thought, big solutions required big research for big projects: fast-breeder reactors, nuclear fusion, etc. The idea that big solutions might come not from high-energy but from solid-state physics; not from a grand technological design but from innumerable small initiatives; not from ever more advanced science but from comparatively simple scientific principles, is felt to be a debasement of the knowledge so far achieved – which, on its own terms, is of course prodigious. It is for just this reason that physicists will not and cannot acknowledge that their activities might be of no use or even detrimental to society.

'You can't put the genie back in the bottle.' Many people believe that once a discovery is made, it will and must be exploited. Yet society has not in fact done everything it has the power to do. Time after time, technological solutions have been marginalized, not pursued or even simply forgotten about – perhaps only to be rediscovered decades later and introduced in combination with other, newer technologies. The history of power generation offer examples of this in abundance: the long-forgotten airship, which is now finally enjoying a renaissance; the electric car, which is an idea that has been around as long as the internal combustion engine; electrolytic hydrogen production and the fuel cell; wind turbines; and many others. The question of what gets researched, developed and introduced depends more on social values and powerful interest groups than on scientific results.

'In precisely this assumption, that scientific knowledge should dictate my actions, lies the greatest hubris, the greatest rape and the blindest error of mankind – for therein lies the destruction of the individual.' So wrote the philosopher of science Viktor Gorgé.[27] According to the natural scientist T von Uexküll, science allows mankind:

> to reconstruct the natural world wherein he must dwell, but he cannot deduce from it the standards by which he must live there as a human being. The standards governing our behaviour towards one another, towards the community and towards ourselves are not dictated by either biology or physics. Indeed, the scientific method does not even reveal to us the standards by which we are to make human application of the possibilities that physics and biology hold forth.[28]

Nevertheless, there has been a succession of eminent scientists, in the 20th century in particular, who have declined to consider the social context of their research, and who thus must share the responsibility for catastrophic consequences and abuses. The core reason, which is cited particularly in connection with nuclear technology, runs as follows: because the acquired knowledge of what is scientifically possible cannot be erased, the only alternative is practical application – and if we don't, somebody else will. Such reasoning quickly turns researchers into allies and aides of those who have a commercial or power political interest in an exploitable scientific discovery.

The history of scientific and technological development over the past two centuries shows a clear link between the quest for knowledge and commercial interest in its exploitation. Today, this link is particularly clear with respect to research on genetics, and it has also shaped the energy industry. The mid-19th century was the era of thermodynamics. From the Carnot Cycle (1815), through the work of James Prescott Joule (1818–1889) and the First Baron Kelvin of Largs (William Thomson, 1824–1907), it culminated in the codification of the Second Law of Thermodynamics. This development occurred almost exactly in parallel with the rise of the steam-engine, whose demand for energy sparked the birth of the energy industry. The energy physics of the time focused on the immediate practical applications but, in so doing, it also narrowed its horizons to the exploitation of fossil fuel energy. At the end of the 19th century, James Clerk Maxwell (1831–1879) developed his comprehensive description of the phenomena of electricity and magnetism, clearing the way for the electricity industry. Maxwell's success was tied to the growing industrial necessity to produce, distribute and apply energy to industrial processes.

At the dawn of the 20th century, the edifice of the physical sciences seemed complete. It was now possible to describe and predict natural phenomena. Industrial development apparently built purely on physical laws was seen as the crowning achievement of physics, which came to be regarded as the queen of the sciences. The age of technical invention had come; the aim was now to continue a chain of development that reflected the fundamental supremacy of the physical sciences. Marie Curie's (1867–1934) discovery of radioactivity in 1898 opened a new chapter, although no-one at the time could have foreseen its consequences. Nuclear physics only properly came of age in 1932, the year in which the neutron was discovered and when the first accelerated chain reaction took place. The second world war resulted in the mobilization of financial resources for atomic research, which lead not only to the development of weapons of mass destruction, but also to the 'peaceful use of nuclear energy' and thus also to the rise of the nuclear industry.

With the advent of nuclear technology, the relationship between energy physicists and the power structures of society became increasingly symbiotic. Researchers might originally have been motivated by a selfless desire to build a better tomorrow; equally, it could have been the egocentric desire to play with big toys and satisfy curiosity, or the dogged quest to secure the respect of fellow scientists and funding for future research. The need for funding grows as technologies get larger, riskier and more complex. As a consequence, scientific and technological research has become ever more susceptible to the powerful interests of various ideologies. The more technology came to shape economic and social development, the more scientists presumptuously sought to control the direction of social change wrought by scientific and technological advances. Many thought that society could be treated as a closed system whose behaviour can be predicted as easily as physical processes.

Energy physicists play a decisive role in this process. Their social standing has always served to undermine projects, or to hold back technological developments that ran contrary to existing trends and established structures. The economic powers-that-be have always deployed scientists and their divergent opinions to suit tactical ends. Joachim Radkau gives examples from the development of the nuclear industry between 1950 and 1970.[29] In the immediate post-war period, scientists appeared to personify the state's new image of progress, prosperity and vision for the future. Little time was needed to complete the ideological leap from the detonation of the first atomic weapons to the myth of the inexhaustible energy source, and for decades, energy physics paid little attention to renewable energy. During the realization phase of a new technology, when the perceived business opportunity mobilizes industrial resources, the role of scientists, and more particularly engineers, is to optimize the new discovery for use. This is when the debate over achievable efficiency takes place: to the scientist, a challenge to prove his or her ideas in the field; to the engineer, an opportunity to build better and more powerful machinery; and to the investor, simply a greater return on capital. The implication of this triangle of interests is that anyone who presumes that an energy physicist embedded in the conventional energy system will be scientifically objective – or even just well informed – when it comes to renewable energy is underestimating the tangled interests of the physical sciences, and overestimating scientists' open-mindedness and awareness of the wider implications of technology. In an appendix to his play *The Physicists*, Friedrich Dürrenmatt writes: 'the subject of physics is the concern of the physicists; the effects the concern of all. What concerns everybody can only be solved by everybody.'

The fear of the small scale

The mythology of the fossil energy industry rests on well-worn experience and habits of thought. The inhabitants of the industrialized countries have had a century to grow accustomed to living in the centralized structures of the fossil energy system. These structures have now become a matter of course, to the extent that most people cannot imagine that energy could be supplied in any other way. The same applies to energy companies and political institutions, just as it does to the general public and to scientists, and often even to proponents of renewable energy. People are accordingly highly sceptical that energy supplies can be guaranteed from renewable sources alone through large numbers of local micropower plants. This scepticism reaches the proportions of an all-out fear of the small scale, fear that small plants might not meet current industrial standards or be able to maintain our standard of living.

The visionary goal of achieving a locally based power supply follows above all in the footsteps of Ernst Fritz Schumacher's dictum that 'small is beautiful'. Yet decentralized structures are not an end in themselves which must be pursued in all circumstances and which must be beneficial in every case. The local approach is not always appropriate, especially as society has come to depend on the low cost of mass-produced consumer goods. It is scarcely conceivable that cities could be supplied with water without a central supply system. Centralization of electricity supplies was government policy in almost all countries, in order to secure an electricity supply independent of the location of the consumer, and to make prices as low and as uniform as possible. There were also environmental reasons, for example that innumerable individual coal and wood fires are more polluting than supplying heat from a central power station. Decentralization is not always desirable.

The problem with centralization is that it became an ideological conviction and was consequently applied to situations where it was actually counterproductive. Large-scale power plants that replace co-generation plants, thus preventing economically and environmentally sensible use of heat; centralized waste disposal schemes that create problems rather than solving them, because of the waste shipments they necessitate: these demonstrate that centralization can be counterproductive. Although if solar power did require large, centralized plant, fundamental environmental considerations and limited fossil resources would make a centralized solar power supply preferable to even a decentralized fossil fuel supply. Solar power, however, only reaches its optimum potential within a decentralized structure, and it is this technological consideration that makes decentralization necessary. At the same time, this means moving from the idea of third-party supply to the autonomous generation of power, towards self-sufficiency and independence. But as broad as the appeal of this approach may be in principle, it has yet to make itself felt in the practicalities of everyday life. Greater autonomy places additional intellectual and practical demands on the individual, whereas the general cultural trend is going the other way, towards the convenience of consuming centrally supplied products.

Political decision-making is also becoming more and more centralized, through an ever denser and more tightly interconnected web of treaties. Consequently, more and more people find themselves confined to the role of consumer, looking on approvingly from the sidelines. If the future demands that people play a more active role again, then many may feel that too much is being asked of them. This makes the return to integrated solar energy systems that overcome the functional division of labour in energy production more difficult.

Nevertheless, there are aspects of society in which social functions have been or are being decentralized. One example is the private car, which has granted greater freedom to large numbers of people, turning them into connoisseurs of automobile technology and forcing a retrenchment of public transport. The car would not have become so popular were it not for the greater individual mobility that it brought; when properly calculated, cars are significantly more expensive than public transport, and purchasing and correctly maintaining a vehicle, like energy systems, demands time and initiative. Another example is IT, which makes individual access to information and its transmission considerably easier and independent of location. People are basically prepared to assume greater responsibility for their daily lives if the result is greater freedom. Unfortunately, both these triumphs of individualization had serious social and environmental consequences. The destructive effects of the car are obvious. In the case of IT, the much-lauded 'paperless office' has resulted in a massive increase in paper consumption, and the practice of ordering goods online from the cheapest supplier will lead to an explosion in the transport industry. Communications technology makes car travel more attractive, producing a further increase in traffic. No one single user intends these effects to happen, but they are accepted as the price of technology. The resultant increase in the consumption of fossil fuel and other natural resources amounts to an irredeemable mortgage on the future.[30]

Putting energy generation in the hands of the people, on the other hand, would have no appreciable negative effects. Individual freedom and collective social responsibility for the future are not mutually exclusive, but rather go hand in hand. What technology could be more desirable than one whose use cleans the environment rather than damaging it? In order for individual energy production to become truly popular, it must also bring tangible advantages in terms of freedom and opportunities, which means the switch must simplify rather than complicate matters.

Solar power is currently still laborious because there has been a lack of suppliers, sources of information and advice, because there are still numerous bureaucratic obstacles, and because the entrepreneurial, technical and personnel infrastructure remains inadequate. Further complications arise because most individual systems are still only partial solutions, which exist alongside nuclear and fossil fuel energy rather than fully replacing them. Most solar collectors installed on houses only supply part of the heating need, and most PV installations only satisfy part of the need for electricity. The operators of such systems thus have to work with both conventional and alternative systems simultaneously, incurring costs for two separate supply systems. Even the personal computer would not have been introduced so quickly and on such a large scale if it had not been able to fully replace and improve on the typewriter.

This is another reason why the fossil energy industry has been able to rely on its claim of irreplaceability. It offers free home delivery, thus concealing the greater complexity of its supply system from the end-user. The more that practical obstacles facing solar energy can be removed, the faster the psychological hurdles will fall – perhaps swifter than was the case with decentralized mobility and IT. Solar technology has no negative environmental consequences to trouble users' consciences. Once the fear of the small scale has been dispelled, once renewable energy has demonstrated that it can replace fossil energy in its entirety, then the aura of the centralized nuclear/fossil energy industry will quickly fade. While fossil mythology remains unchallenged, humanity is faced with the absurd prospect of choosing death over a solution it is afraid to embrace.

Yet the man on the Clapham omnibus is only afraid of the alternative because the great and the good tell him to be. Ordinary people, above all, are caught up in the myth of big technology, for many of the reasons already mentioned: the assumption that big solutions mean big technology; backing technocratic implementation over an active and engaged society; lack of strategic imagination; overblown and uncritical respect for the achievements of science and technology, however problematic they may be; the cowardice of political institutions in the face of the large energy corporations or the all-too-close interconnections between the two. Anxiety in the face of an alternative requiring innumerable small steps has been implanted into the public consciousness by those who really have something to fear, because the interests of the energy industry are threatened. The latter's fear is real; the public's, by comparison, is illusory. It is a product of the mythology surrounding the energy industry. It is time for this mythology to be unmasked.

PART III

THROWING OFF THE FOSSIL SUPPLY CHAINS

One game that energy economists like to play is tracing the growth in market share over time of a particular energy source. The curve always climbs slowly when a new energy source is first introduced, but then it takes off – before gradually falling back towards zero. The insight that these curves are supposed to give is banal. Obviously it normally takes time for a new energy source to be introduced, because of the inevitable industrial lead times. What is not banal is when such curves are supposed to give the idea that renewable energy cannot and should not be rushed onto the market.

Market analyses distract from the fundamental differences between renewable and fossil fuel and nuclear energy sources. The two are simply not comparable. Solar resources can be deployed faster than previous experience would suggest, because unlike fossil fuels, there is no need for a complex supply chain. Solar resources have the potential for extremely rapid deployment, providing that technology and business strategies are geared up to cope with the unique opportunities they offer.

The technological pioneers of the industrial revolution and the companies who built the modern technological world – Edison, Siemens, Bosch, Daimler, Ford and others – were exploring virgin territory. Though they may well have faced the same resistance, disbelief, prejudice and fear of change, they did not have to contend with opposition from established, powerful industries. It is a highly ambivalent situation: under these conditions, technical innovations that promise to unleash a new wave of development can take off faster than ever, providing they fit entrenched interests and offer large corporations opportunities to consolidate and expand their markets. If, however, innovations run counter to or even endanger corporate business models, then they will be opposed with the massed might of tightly knit business structures. Large corporations do not look kindly on what they see as trespassers on their territory.

Of all the cases of radical technological innovation analysed by James M Utterback of the Massachusetts Institute of Technology, only one quarter were championed by established companies. Older companies prefer to spend hundreds of millions improving on established products (so-called incremental change) than one or two million on developing new products for which the existence of a market has not been proven. The market seems 'too small', the risk 'too high'. Established firms have no interest in relinquishing proven technology and/or established markets. They timidly assume that the new product will take too long to penetrate the market, if it ever does. For large companies, only the certainty of expectations or experience can justify a change of strategy or the ploughing of a new furrow. For this reason, either they do not engage at all with radically new technologies that replace existing products, or they do so only half-heartedly. Yet according to Utterback, 'total engagement' is needed to make a new technology work.[1] If on top of this, as in the case of energy, existing firms are also fettered by established supply chains, then the tentative approach large companies are taking to renewable energy becomes easier to comprehend.

For precisely this reason, it is of central importance that the technologies for converting and using solar energy should be conceived in such a way that, like the steam-engine before them, they will become an unstoppable economic force. The $64,000 question is: what are the 'killer applications' for solar resources?

6

Energy beyond the grid

ACHIEVING MORE EFFECTIVE and more comprehensive energy supplies has always meant the construction and expansion of energy grids. Energy grids have become bywords for economic progress and prosperity. The advent of grids, however, heralded the seemingly irrevocable demise of autonomous energy production. The wider the reach of the distribution grid, the larger the suppliers, and vice versa – irrespective of the commodity supplied, be it electricity, gas, water or heat. The same also goes for the oil and coal distribution networks. The national grid for electricity in particular symbolized the conclusion and perfection of the modern energy system. Consequently, electricity generation technologies are evaluated and selected on the basis of their compatibility with the national grid, and the technologies for generation from renewable sources are also being developed for seamless integration with the grid.

Technologies for autonomous power generation, ie, with no grid connection, are not taken seriously by energy experts. They are at best regarded as special cases, makeshift or childish nonsense, appropriate only for niche applications or backward regions of the developing world. Energy-hungry society usually regards those who – for idealistic reasons – aim for complete energy autonomy as cranks and oddballs. Energy-autonomous living, including an array of autonomous devices and solar-powered vehicles, was all the rage in the USA of the 1970s, drawing on the ideals of individual freedom of the civil rights movement.[1] Mainstream society jeered or simply took no notice. Efforts to develop solar power technology have consequently given little consideration to the idea of autonomous power supplies (albeit with the exception of PV installations for developing country villages with no grid connection). This focus on integrating renewable energy into the national grid shows up most clearly in the lack of research into electricity storage.

In fact, it is the unique capacity for autonomous and local-grid power supplies that only renewable energy can offer which presents the greatest opportunity to break energy supply chains and revolutionize economic structures. The following sections consider the technological options available.

Wireless power: the potential of solar stand-alone and stand-by technologies

Grid-independent PV is already ubiquitous. It started with pocket calculators, solar wristwatches, portable radios and miniature pumps for back gardens or signal buoys, which draw their power from built-in solar cells. The range of devices available is continually expanding, and most can be ordered through appropriate catalogues. The most prominent distributors are Real Goods of California and the solar technology mail-order company GWU, based in Fürth, Germany.[2] The variety of autonomous technologies ranges from solar-powered road-sign lighting, parking meters, electric fences, electric razors, cameras, hand-drills, automatic garage doors, emergency telephones, lamps, lawn-mowers, hand-held vacuum cleaners, ventilators, detectors for domestic alarm systems, automatic teller machines (ATMs) and street lighting through to in-car air conditioning powered by solar sun-roofs, and many others. In the 1980s, the German research ministry even sponsored a research and development programme for small devices like these at the Fraunhofer Institute for Solar Energy Systems in Freiburg, under the leadership of Adolf Götzberger and Jürgen Schmidt, although the programme was later scaled back.[3]

Even so, the ranks of solar-powered stand-alone devices have long since included more demanding applications, such as battery chargers or rechargers for mobile phones; mobile phones and powerbooks containing built-in solar panels; light-houses and radio beacons; and boats and refrigerated lorries powered by solar cells alone, the latter drawing power for the refrigeration system from the solar panels on the trailer roof. The possibilities are almost unlimited, potentially including all devices that currently run on mains electricity or batteries: desk or standard lamps in closed but well-lit rooms, remote control units, the whole gamut of household appliances through to fridges where the door could be a solar panel. Each individual application could be dismissed as a side issue, but in sum, providing that they become mass-produced standard appliances, they could quickly become important.

Electricity companies were never generous enough to underestimate the potential impact of household and office gadgets on their sales. The availability of a wide array of specialist gadgets enhances turnover. The appreciable rise in electricity consumption in homes and offices over recent decades can largely be attributed to innumerable labour-saving devices and the systematic electrification of everyday household and office equipment. In Germany, homes and offices consume 200 billion kWh annually, which is 38 per cent of the entire electricity supply.

The term 'stand-alone system' covers both those which are wired up and always ready for use, and those which function or can function completely independently, without wires. Take the apparently small example of the household doorbell. The transformer for a doorbell consumes between 9 and 22 kWh a year. For the around 37 million households in Germany, that adds up to a total consumption of over 500 million kWh annually, equivalent to the electricity demand of a town of 100,000 inhabitants. A single, matchbox-sized PV module mounted on the wall by the bell would be enough to keep the bell going. To put it another way, this would equate to the installation of 500 MW of photovoltaics, four times the annual world output in 1998. Moreover, a stand-alone doorbell would not need a transformer and, in detached houses, there would be no need for additional wiring. The result would be cheaper or at least as cheap as conventional doorbells.

Many stand-alone devices are powered by batteries, both rechargeables and non-rechargeables. In 1997, turnover in the global market was $35 billion, and the industry is expecting growth of 5 per cent per annum, resulting from the demand for consumer electronics in particular.[4]

Most batteries, both rechargeables (and accumulators) and non-rechargeables, could be replaced by built-in solar panels or solar-powered chargers. This would even make the devices more user-friendly, as they could be recharged on the hoof without needing to find a wall socket. In the case of solar pocket calculators, for example, there would also be no need to worry about changing non-rechargeable batteries. Non-rechargeable batteries would largely disappear from the market. Mobile phones that can recharge in any available sunlight would save users from worrying whether the power will run out at an inopportune moment; likewise, laptops with solar cells in the lid could recharge during use. Mobile phones currently need 35 kWh a year on average; base-station rechargers for cordless phones need 42 kWh. Making these devices autonomous using solar panels would save the individual user around seven or eight pounds a year. Twelve million such devices – the estimated size of the mobile phone market in Germany – would represent the replacement of 900 million kWh of conventional electricity with solar energy. The substitution of primary energy would be three times as high, once the various losses along the supply chain have been factored in.

Since solar panels can be incorporated into every electronic device, constant improvements in the conversion efficiency of solar cells, the energy efficiency of the devices and the capacity of accumulators offer ever greater opportunities to ensure that the growing number of electronic devices does not translate into greater demand for mains electricity. 1978 saw the USA's first grand plan for industrial production of PV. But although it received authorization, it was never actually implemented. The entire army was to be equipped with solar-power field telephones, in order to eliminate the need for batteries – even back then, it had been calculated that this would cost less than the conventional power supplies of the time.[5] Furthermore, solar cells are not an issue with respect to waste disposal – unlike batteries, which present a serious toxic waste problem.

The usual cost comparisons per kWh do not apply to built-in solar panels, because this would no more be an issue for people than it is with batteries. The amount of electricity consumed by battery chargers cannot be measured statistically. The necessary charging time is usually exceeded, sometimes by hours or even days; the resultant loss of electricity is pure waste. As a result, mobile phones, for example, probably consume far more power than they actually need. Accumulators and the transformers that feed them also waste significant quantities of electricity. When devices are not used for an extended period of time, the loss of electricity due to self-discharge is considerable, up to 95 per cent of the actual amount used. Transformers also consume electricity in converting 240 volts (V) mains to levels usual for electrical devices of between 1.5 and 60 V, and they do this even when the devices they supply are switched off.

Consumer surveys have been used to calculate moderately reliable figures for the energy wasted by the mains-backed stand-by modes of televisions, video cassette recorders (VCRs), hi-fis, fax machines, hot water boilers, household appliances with built-in clocks, telephone extensions, answering machines, CD players, and personal computers (PCs) with monitors or modems in the home and at the office. If all households each ran only one television, one satellite receiver, one VCR, one answering machine, one hi-fi and one fax machine, then using figures worked out for average stand-by power consumption, this adds up to an annual electricity demand of almost 600 kWh, at an additional individual cost of €62 ($55) a year, for household electronics in stand-by mode. It is calculated that Germany wastes 20 billion kWh annually on stand-by functions, at a cost of over €2 billion, or nearly $2 billion. This is the same as the electricity consumption of Hamburg, Berlin, Munich and Frankfurt combined. For comparison, statistically measured electricity production from renewable sources in Germany in 1998 was around 25 billion kWh. For the EU, electricity consumption by devices on stand-by has been calculated to be 100 billion kWh a year, or one fifth of Germany's total electricity consumption of 500 billion kWh per annum, equivalent to a conventional power station capacity of around 20,000 MW. These figures were reached even without accounting for the losses from inefficiencies within electronic devices. In most cases, power is supplied to all parts of an electronic device even when only one component is being actively used, rather as if a house only had one light switch, thus making it impossible to light only one room at a time.

Stand-by functions are now the subject of heated debate. Demands for new, energy-efficient appliances meet with the rejoinder that stand-by mode is an unavoidable necessity for devices like answering machines or fax machines, which have to be 'always-on' to work.[6] Considerable development effort goes into reducing the power consumption of these appliances, and thereby reducing or avoiding the losses due to always-on operation. There are awareness-raising advertising campaigns and specialist workshops aimed at producers, sellers and customers, labelling regimes mooted, and so on. Yet hardly anyone is making the case for what would be the most obvious solution, namely that all these problems could be easily circumvented by building solar panels into always-on appliances.

If stand-alone and stand-by functions were to be powered in future primarily by solar panels – in the name of lower costs, greater convenience and environmental protection – this alone would probably raise the share of renewable energy to well over 10 per cent of the total electricity demand, enough to replace at least 10,000 MW of capacity in Germany alone. Around ten times the current world output of PV would be needed, which could give the technology such a boost that it would quickly become faster and cheaper to deploy for other, larger-scale applications. As the power for always-on devices forms part of the base load, allegedly a no-go area for PV, such solar-powered devices would also effectively demolish one of the standard arguments against solar electricity.

This presents an opportunity of an order of magnitude that promoters of PV scarcely dare envisage for the foreseeable future. Industrial suppliers will need imaginative product development and marketing concepts if they are to rise to the challenge of bringing out superior appliances and making them succeed in the marketplace. For makers of conventional always-on and stand-alone devices, the effort required would be no more than improving on established, saleable products, and no great leap into the unknown. For the electricity industry, of course, it would mean a painful market contraction.

The opportunities for cordless power just described also demonstrate how solar and energy-efficient technologies can complement each other in a productive way, rather than being treated as mutually exclusive. For the greater the level of energy efficiency, the faster solar power can grow – and help erode the interwoven structures of the conventional energy industry. Hans-Joachim Bruch, an engineer and advisor to the German environment ministry, calculated for this book how large and how powerful a built-in solar panel would have to be in order to power the stand-by mode of an always-on device the long way round, through the mains (see Table 6.1). As solar panels could supply power not just when the device is inactive, but also while it is in use, solar-powered always-on devices could replace double the 20 billion kWh that stand-by functions are calculated to consume, ie 40 billion kWh. Including losses over the entire electricity supply chain, that adds up to the replacement of a total of 120 billion kWh of primary energy!

There are no economic factors holding this development back; only the lack of imagination on the part of established industries and the opposing interests of the battery and electricity companies. That global producers of electrical devices such as Bosch have largely closed their minds towards this potential for (solar) technological innovation does not exactly show them to be forward-thinking companies.

The situation in Japan is already rather different. The government-initiated 'Sunshine' project has secured the collaboration of almost all large companies in the electrical and glass industries and, besides government money, these firms are investing far more of their own resources in the development of PV than companies in other countries – and have been doing so for years. Japanese industry applied for over 6000 patents on PV technology between 1981 and 1995 alone, many of which were for small devices.[7] Apart from a few exceptions like Siemens and Pilkington, European electrical and glass manufacturers have not yet woken up to this issue. Phillips has even called a halt to its initial foray into PV.

Table 6.1 *Stand-by power consumption and equivalent PV panel area*

	Stand-by power consumption		
	Average for existing appliances (Watts)	New, energy-efficient appliances (Watts)	Highest power consumption[1] (Watts)
Entertainment electronics			
Televisions	12	0.1	20
VCRs	15	1.0	28
Satellite receivers	20	3.0	35
Hi-fi systems	12	1.0	14.5
CD players	6	0.1	7
Household appliances			
Electric cooker with built-in clock	6	3.0	7
Microwave oven with built-in clock	3	3.0	4
Coffee machine with built-in clock	4	2.0	5
Communications equipment			
Telephones (2–10 units)	20	8.0	25
Answering machine	4	1.6	12
Fax machine	12	1.0	100
Home computers			
PC and monitor	100	2.5	200
Ink-jet printer	10	2.0	70
Modem	8	3.3	10

1 Highest power consumption: primarily older units.
2 Output in watts per module (with optimum orientation) for equivalent annual
power consumption in stand-by mode; panel area needed in m^2, assuming
11 per cent efficiency without generation and transmission losses.
Source: UBA/Hans-Joachim Bruch

Annual stand-by power consumption			PV panel needed to supply the appliance, assuming that the power is transmitted over the grid (output and area)[2]					
Average for existing appliances	New, energy-efficient appliances	Highest power consumption[1]	Average for existing appliances		New, energy-efficient appliances		Highest power consumption[1]	
(kWh/yr)	(kWh/yr)	(kWh/yr)	W	m²	W	m²	W	m²
83	1	139	95	0.87	1.2	0.01	160	1.45
126	8	235	145	1.32	9.2	0.08	270	2.45
139	21	242	160	1.45	24.1	0.22	278	2.53
96	8	116	110	1.00	9.2	0.08	133	1.21
50	1	59	57	0.52	1.2	0.01	68	0.62
48	24	56	55	0.50	27.8	0.25	64	0.59
26	26	35	30	0.27	30.0	0.27	40	0.37
12	6	15	14	0.13	6.9	0.06	17	0.16
161	4	200	185	1.68	73.6	0.67	230	2.09
35	14	104	40	0.37	16.1	0.15	120	1.09
104	9	870	120	1.09	10.3	0.09	1000	9.09
44	1	88	51	0.46	1.2	0.01	101	0.92
4	1	31	5	0.04	1.2	0.01	36	0.32
2	1	2	2	0.02	1.2	0.01	2	0.02

The potential for natural and technological solar energy storage

The lack of an adequate mechanism for storing electricity is the greatest handicap faced by PV and wind power. The current solution is to use the grid as a form of proxy storage: renewable energy is fed into the grid; when there is no wind or sun, conventionally generated electricity is supplied from the grid.

Operators of conventional power stations argue against this on the basis that the unreliability of wind and sun means that total operating time over the year is so low that economic production from renewable sources is not possible, even with a large number of PV and wind power installations feeding power into the grid. Peak operating time for well-situated windfarms is around 2000 hours in a year of 8760 hours, albeit this is increasing as the technology improves; for PV, peak operating time is less than 1000 hours. In reality, installations run for longer than this – around 4000 hours in the case of wind – but at reduced output. This puts into perspective the electricity industry's plea that, while they have to have capacity in place to cover for wind and PV downtime, it languishes unused when electricity is being generated from these sources. For as long as wind and PV output is a negligible proportion of total production, this is a flimsy argument. Electricity producers have so much underused capacity.

Nevertheless, 'capacity effects' do come into play when a large proportion of energy is generated from wind and PV. In this case, there would be a need for reserve capacity that would indeed be underused during wind and PV operating hours. Targets for increasing the share of electricity from wind and PV are thus sowing the seeds for future capacity conflicts between operators of PV and wind installations and the conventional electricity suppliers. Consequently, the industry is demanding that feed-in legislation for renewable energy be dropped, or upper limits imposed on the 'forced purchase' of renewable energy.

One possibility for postponing the capacity conflict between PV and wind and conventional power stations would be to expand the scope of the grid. There is, however, an internal contradiction in the idea that the economic viability of decentralized plant should depend on a large energy grid, of all things. It also makes electricity from PV and wind unnecessarily expensive, as transmission and distribution make up around 60–80 per cent of electricity costs. The only reason that this is not more obvious is that electricity bills make no mention of it, and the electricity industry does not publish any precise statistics on the subject.

If the economic dynamism latent in PV and wind power is ever to blossom, new answers must be found for two key questions:

1   How can the need for expensive reserve capacity be avoided in future?
2   How can the unique economic advantage of renewable energy sources, in that their exploitation does not require long supply chains or an overarching energy supply system, be made to count for wind and PV power in practice? To put it more tendentiously: how can they be made independent of anonymous reserve capacity and even independent of high-performance energy grids?

Hybrid systems: electricity supply without anonymous reserves

The first point to note is that reserve power stations lose their significance if the grid is fed from numerous small local power stations. Unlike large power stations, it does not matter if a small power station drops out. A system built of large numbers of small plants is inherently multiply redundant. Secondly, continuity of supply is not an argument that sensibly applies to a market economy in which the value of a good is determined by the interaction of supply and demand. Thirdly, the correct response to the argument that you can't dry your laundry if the wind doesn't blow is that you can dry it when it does blow.

One suggestion for solving the problem would be so-called hybrid systems that can generate electricity from two different sources – and above all harnessing all known and as-yet unknown technologies for storing electricity once generated. With hydropower from reservoirs, it is of course possible to meet all electricity demand from one source, since it is possible to react immediately to changes in demand by swiftly controlling the water flow to turn turbines on or off. Where there are enough suitable sites, as in Norway, the entire national demand can easily be met from hydropower stations. Many smaller regions could also make themselves independent of external electricity suppliers. Yet they choose rather to sell their hydroelectricity far and wide. Its suitability for meeting peak demand commands a high price on the energy market. Any further source of electricity can supply an electricity market's needs if combined with sufficient hydroelectric capacity, be it fossil fuel or renewable. France has a hybrid system that combines hydropower with nuclear energy in the ratio of 1:3. A strategy that works at national level can quite clearly meet all electricity needs on the small scale – without nuclear power and without fossil fuels, but also without the universal availability of hydrolectricity from reservoirs.

Another possible hybrid would be wind power in combination with a biomass plant. Whenever the wind dropped, but demand for electricity remained high, a biogas-, vegetable oil- or gasified biomass-burning generator linked to the windfarm would automatically start up, and stop again as soon as the windfarm took the strain again. A plant like this could supply the grid according to need, or could guarantee a fully autonomous power supply. The argument that a fossil-fuel-free energy supply cannot be achieved without recourse to geographically limited reserves of hydroelectricity quite literally does not hold water. There are other reasons why such a hybrid supply is not the only and not even the optimum strategy for renewable energy. A biomass-fired generator serving as reserve capacity for a wind farm would not be fully utilized, since it could be providing power round the clock. Furthermore, the primary energy could be used most efficiently if the waste heat were captured and put to use as well. The traditional seasonal differences in energy usage, however, make it scarcely possible to guarantee custom for both heat and electricity if the generator is run at constant capacity. All these factors make the question of direct storage of electricity a pressing issue.

Technologies for storing solar energy

In 1999, the Society for Innovative Energy Generation and Storage (EUS) commissioned an electricity storage plant in Bocholt, Germany. This plant stores electricity from four wind turbines with a total output of 3.5 MW in a 1.6 MW battery, and feeds it into the grid whenever demand is greatest. Due to the greater returns enjoyed by this plant, it is expected to have paid for itself in around six years, despite being the first installation of its kind.[8] This and other storage technologies open the door to far-reaching economic opportunities for renewable energy to blossom in the electricity market. Once it becomes possible to store electricity, all the arguments against supplying electricity from renewable sources relating to capacity and productivity lose their sway, as do previous arguments for the existence of a national grid. Low-cost storage technologies enable a qualitative leap to exploiting renewable energy across the entire electricity supply system. Ultimately, decentralization of energy supplies will become unstoppable.

The spectrum of potentially useful storage technologies ranges from electrochemical, electrostatic and electromechanical to thermal and chemical media. Most widespread so far has been electrochemical storage, in the form of batteries. There industry let it rest for a long time, in the absence of perceptible demand for other, better options. There were many reasons for this status quo. Though the idea of electric vehicles had been around for a century, so few were built that there was no pressure on industry to develop lighter, more powerful batteries. Regional monopolies also meant that electricity companies had no incentive to devise new storage media to cover peak demand, for example: instead, they preferred to rely on dams, reservoirs and pumps. Only for submarines were more powerful lead-acid batteries developed. These subsequently came to dominate the industry.

The most effective pressure to develop new storage technologies in the past two decades has come from the environmental movement, which demanded less polluting batteries, and from the electronics and space industries. The latter's need for tiny mains-independent devices with an

extremely low power rating led to the development of electrostatic supercapacitors, which are also vital for the development
of solar-powered stand-alone and always-on devices as outlined
earlier. Recently, the car industry has found a need for powerful batteries to optimize the technology for electric cars, for
which the greatest spur so far has been California's environmental legislation. This requires 10 per cent of all cars sold in
the state to be zero-rated for emissions by 2003. In the mid-
1990s, the US government also initiated a $260 million
research programme in battery technology, in which all US car
manufacturers are participating.[9]

Electricity storage has up to now played only a secondary
role in publicly funded research into solar technology, despite
its central importance for renewable energy supplies. Hence it
is important to be aware of the whole spectrum of options
available. Given the decades-long neglect of electricity storage
technology, none of these options can have had time to mature;
nevertheless, the range of possibilities is broader than many
people realize, and it is possible in some cases to draw on tried
and tested technology, which, in combination with renewable
energy, can now be put to new and undreamed-of uses.

Electrochemical accumulators

In electrochemical batteries, power flows in through one
electrode and out through a second. In the process, the energy
content of the chemical substance between the electrodes is
increased. The process is reversible, and can be repeated
thousands of times. The commonest form of electrochemical
accumulator is the flooded or hermetically sealed lead-acid
battery, which is now largely a mature technology. They are
cost-effective and highly efficient, but have a low energy
density, and their disposal causes considerable problems. Better
energy densities, albeit with lower efficiency, are offered by
nickel-metal-hydride batteries; but here too there are problems
with disposal.

One new type is the redox battery (short for 'reduction and
oxidation electrolyte circulation'), which has a viscous fluid
electrolyte. Once drained of charge, the fluid is pumped out at

a garage, and replaced with charged fluid. Changing the electrolyte fluid saves the user the hours required to recharge the battery, thus making them particularly suitable for use in electric cars. Efforts to drastically reduce battery weight are also directed at electric cars: lighter batteries would extend their range.

More promising for the issue in question, however, are lithium ion or lithium polymer batteries, which take the form of a thin film. The technology is still very new, and the costs correspondingly high. But efficiency and energy density are high, weight is negligible, and the batteries are good for innumerable cycles, environmentally sound and require virtually no maintenance. Lithium ion batteries also do not need special chargers. They make particular sense for PV because the batteries can be built into the panels, thereby integrating generation and storage in one unit. Building roofs and façades would also be suitable storage surfaces.

Electrostatic storage

Supercapacitors come into this category. Electricity is stored without loss in a solid electrolyte, and no chemical change takes place. Supercapacitors are light and can be extremely small. Though still immature, the technology combines high energy density and efficiency with low environmental impact. Their working lifetime is greater than for all other battery types, stretching into the millions of charge/discharge cycles. The cost, though, is still high, and current models are not very powerful, having been developed for low-power electronics. The first supercapacitors could store no more than a few ampere-seconds; this has since been increased to an ampere-hour. Currently to be found in wristwatches, mini-radios and measuring instruments, supercapacitors are vital to the technical development of stand-alone and stand-by devices. They offer considerable additional scope for reducing energy use in all electrical devices, thereby smoothing the way for cost-effective growth in PV.

Fly-wheels

Fly-wheels are a form of electromechanical storage. The fly-wheel is a rotating cylindrical body, and the amount of energy stored grows in proportion to the mass of the fly-wheel and the square of its speed of rotation. The stored energy can be use to drive a motor or to smooth out short-term fluctuations in the supply or flow of energy. Fly-wheel technology has a variety of applications from cassette recorders to motorbikes, but also including motor vehicles and motor generators. Years of neglect have left the technology underdeveloped, but energy densities are high and there is no waste disposal problem.

Researchers are currently experimenting with magnetic fields as a means of achieving higher speeds and reducing the loss of stored energy due to friction caused by the weight of the spinning mass. An electrical motor and braking system is used to control the speed of the fly-wheel, which currently reaches around 120,000 revolutions per minute. Fly-wheels are easy to manage and can be scaled down, which makes them suitable for local autonomous supplies, to bridge gaps in supply from wind or PV.

Compressed air

Compressed air is a tried and tested technology that can quickly be deployed for electricity storage. Factories used to drive their machinery with it; now compressed air is used to enhance the performance of Formula 1 and aircraft engines. Compressed air is another form of electromechanical storage. Electrical energy drives air compressors which pump air into high-pressure tanks. The stored air is then used to drive generators or motors as required. Compressed-air tanks are well understood, the costs are relatively low, and energy density average.

The first public demonstration of a car that runs solely on compressed air took place in 1999, the result of a collaboration between former Formula 1 motor engineer Guy Negre and the Luxembourg firm MDI. The car needs 20 kWh of electricity to fill a 300 litre tank, which gives it an urban range of 200 km (125 miles). The car's top speed is 110 km/h (70 mph).

With electricity costing €0.10 ($0.09) per kWh, it is possible to travel 200 km for €2 ($1.80). The motor draws in normal atmospheric air, and this is fed along with the compressed air into a cylindrical chamber. The heat expansion of both air inputs drives pistons in a neighbouring cylinder, which in turn drive the car. The entire two-cylinder motor weighs only 35 kg. The relatively low weight of the compressed air tank is a considerable advantage over today's electric vehicles, whose range is limited by their very heavy batteries. Using external compressors, for example at a compressed air station, 'refuelling' takes only three minutes; with the on-board compressor plugged into the mains, it takes four hours. Servicing is supposed to be necessary only every 100,000 km; the only emission is cold air. The efficiency is 85 per cent of the electricity needed to run the compressor.

Compressed air does not just make sense for cars. Fixed plant for energy storage in buildings, for example, is even easier to manage. A 15,000 litre tank, comparable to a moderately large domestic oil tank, could store 1000 kWh that could be converted into electricity with the help of a motor. That would be sufficient for an autonomous electricity supply, something that could previously only be achieved with a cellarful of batteries, each with an operational lifespan of only up to 2000 charge/discharge cycles. The number of compress/decompress cycles achievable with a compressed air tank is in principle unlimited, which means that small autonomous units can get by without a grid connection. If building-mounted PV panels are supplemented with a small wind turbine, with a capacity of perhaps 1 kW, or a small Stirling engine, so that the pressure tank can be topped up even when the sun does not shine, then the tank need be no bigger than an average domestic oil tank. It would also be possible to turn wind turbines into compressed air stations, or to combine wind turbines with large compressed air motors to provide a truly 24-hour electricity supply.

Electrodynamic storage

The medium in this case is an electromagnet, like the spark plugs in a car engine. Current is piped into a superconducting coil to create a magnetic field, from which current is subsequently drawn. However, the technology is still very much in the experimental phase. The superconducting coils have to be cooled to 170 degrees absolute, and the relationship between input and output for the stored electricity is still unclear. Moreover, the system is highly complex and heavy.

Solar-powered electrolysis

The power storage option that offers the widest variety of applications is electrolytic extraction of hydrogen, by which electrical energy is converted into chemical energy. Electrolysis is a long-established process; the primary focus of development work is improved efficiency. The electrolysis equipment consists of a cathode (the negative electrode) and an anode with a water-based electrolyte in between. Electrons are forced out of the cathode into the electrolyte, and the resulting chemical reaction releases hydrogen. The anode (positive electrode) sucks electrons out, causing a second reaction which releases oxygen. It is vital to keep the hydrogen and oxygen gases separate. Hydrogen has a high energy density, and therefore requires little storage space. Its extreme versatility makes it the ideal fuel.[10]

What matters is how the hydrogen is produced. If the electricity used comes from nuclear or fossil-fuel power plants, the result is environmental self-deception. Although the fuel is clean – the only emission being the water vapour produced by combusting hydrogen in oxygen – no substitution of nuclear or fossil fuel energy takes place.

The overwhelming majority of schemes for hydrogen production from renewable energy envisage using large power stations – large dams or solar thermal plants in arid and semi-arid regions – to mass-produce hydrogen for subsequent delivery to the end-user. The other option for solar-powered hydrogen electrolysis would be a locally based approach, using electricity from PV or wind. Arguments in favour of this route relate to the opportunities for autonomous fuel production or for storing self-generated electricity rather than feeding it into the grid.

Thermal storage

The question of how better storage options can ensure or further develop energy self-sufficiency is also germane to solar heating. It goes without saying that every solar collector needs a hot water storage tank. The pattern thus far has been, however, that solar heating provides only one part of the necessary heat, with additional heating needs being met from conventional sources. The natural next step is therefore to seek complete independence from fossil fuel top-up supplies. From a technological point of view, this does not present a problem: it just takes a larger collector area, greater storage capacity and less need for heat – for example, with better insulation, heat exchangers, heat reclamation and optimal passive solar gain for the building as a whole. This line of attack has led to many successful zero-emission housing projects, recently even including some by mainstream developers.

But enlarged collector area and greater storage capacity is not the only answer. Another option is the solar magnesium hybrid storage system developed by Hans and Jürgen Kleinwächter in cooperation with the Max Planck Institutes in Müllheim and the Ruhr. The system works by using mirrors to concentrate heat on the storage unit, where the heat energy separates hydrogen from magnesium. The hydrogen can then be used as a heat-transport medium to drive a Stirling engine producing electricity and hot water for the heating system. Once the hydrogen has recombined with the magnesium, the cycle can begin again.[11]

Thermal plants could also do with making better use of seasonal variations in temperature. This is particularly relevant in the case of CHP, or cogeneration, which currently still largely runs on fossil energy. The problem with CHP plants is that as the cogeneration sector grows, it becomes increasingly difficult to find customers for the spare heat. It would make more sense to use the heat to drive Stirling engines producing additional electricity. This electricity can then either be consumed immediately, or stored for later use, as described above. Stirling engines are thermal power plants which do not need a fixed fuel input, being able to convert any external heat source into mechanical or electrical energy.[12]

Pre- vs post-conversion storage

Table 6.2 summarizes the energy storage media just discussed, and offers a crude evaluation of each with respect to the state of technological development, efficiency, energy density (a proxy measure of the space required), maximum number of charge/discharge cycles and the resultant overall performance and environmental impact. The ability to store electricity makes independence of the grid and its controlling influence on electricity supplies a real possibility. It also permits rapid expansion in the amount of conventional energy replaced by PV, wind and solar heating.

No energy supply system can do without energy storage. In the case of fossil fuels, the energy stores are stockpiles of coal and oil and gas tanks. Biomass being just as easy to store as fossil primary energy, storage would not be a particular issue for renewable energy if in future it were all supplied from biomass. However, as Chapter 2 concluded, though this would be possible in theory, it would be completely unnecessary in practice, and there are numerous reasons why it cannot be recommended: the role of biomass as an industrial raw material is crucial to achieving a sustainable economy.

Switching to biomass fuel would prevent further accumulation of greenhouse gases in the atmosphere. But fossil fuel pyromania has so overloaded the atmosphere with CO_2 that more is needed than simply stabilizing atmospheric CO_2 at current levels: it must be actively reduced. Plants absorb CO_2 as they grow, sequestering the carbon and supplying the environment with oxygen. Only with their aid will it be possible to start taking CO_2 back out of the atmosphere. This means that existing plant cover must be increased, primarily through afforestation, such that plants sequester more carbon and release more oxygen than additional CO_2 is emitted into the atmosphere. Burning and regrowing equal amounts of biomass has a net zero effect on the atmosphere. Yet while – unlike fossil fuel consumption – this does not makes things worse, neither does it make them any better.

For any given practical application of renewable energy, it makes sense to choose the form of energy that is most conve-

Table 6.2 *Energy storage technologies*

Storage system	Level of development	Efficiency	Cost	Energy density (footprint)	Charge/discharge cycles	Environmental impact	Note
Electrochemical							
Lead-acid batteries							
flooded	3	3	3	1	1000	1	most widespread
gas-tight	3	3	2.5	1	1500	1	
NiCd	3	2.5	3	2	2000	1	
NiMH	2	2	2	2	1500	3	in common use
Li ion	2	3	1	3	5000	3	in the pipeline/promising
Electrostatic							
Supercapacitors	1	4	2	3	unlimited	3	promising
Electromechanical							
Fly-wheels	2	3	3	2.5	>10,000	3	promising
Compressed air	3	3	4	2	unlimited	3	very promising
Electrodynamic							
Superconducting magnets	1	3	?	1	unlimited	3	application questionable
Chemical							
Hydrogen	2	2	2	4	limited by fuel cell lifetime	3	multiple applications
Thermal/thermochemical							
Hot-water storage tanks	3	3	2	2	unlimited	3	most widespread
Magnesium hybrid	3	3	2	3	unlimited	3	promising

Scale: 1 = poor; 2 = fairly good; 3 = good; 4 = exceptionally good.
Source: Hermann Scheer, Heinz Ossenbrink

nient to store, as far as storage is necessary. To this end, it is useful to know whether the energy is stored before or after the conversion process. The former is the typical pattern for energy derived from fossil fuel combustion. Of the renewable energy sources, only biomass can be stored in its original state. Post-conversion storage is the typical pattern for non-combustive renewable energy generation: direct conversion of sunlight into electricity, wind power, non-reservoir water power or direct solar heating.

Biomass is thus best suited to those applications which require pre-conversion storage, primarily machinery that works on the combustion principle, motor vehicles in particular. Another candidate would be solar hydrogen. The advantages of biomass over solar hydrogen, however, cannot be overlooked. Hydrogen fuel can only be supplied in quantities large enough for the mass transportation market if produced on an industrial scale. The focus of attention in this respect has been on the fuel cell.[13] Fuel cells are also the subject of keen interest from the car manufacturers, who are finally beginning to prepare for a post-oil world. The difference between the solar-electric car as discussed earlier and a car powered by fuel cells is that in the latter case, the electricity is generated on board, rather than taken from a battery.

Fuel cells are power plants in which a form of cold combustion process converts hydrogen or another gaseous fuel into electricity – effectively reverse electrolysis. In a car, this process takes place in the vehicle itself, and the electricity produced is used to power an electric motor. Fuel cell cars should enter mass-market production in the first decade of the 21st century. They will be almost completely silent, and emit nothing more harmful than water vapour. Until hydrogen starts being produced in sufficient quantities, the idea is to run them on natural gas.

Will the need for large quantities of fuel mean that fuel cell cars remain dependent on international energy supply chains – not necessarily on the grid, but on global suppliers of solar hydrogen? Will the motors whose thirst for fuel sparked the globalization of the energy industry in the 20th century be the factor that ensures the continued need for a global energy industry, in the form of the fuel cell and its need for hydrogen produced using electricity from large-scale power plants?

The range of renewable energy options is wider than this, even in the case of road fuel, and irrespective of the scope for fitting cars with more effective and lighter systems for direct electricity storage. Hydrogen can also be produced from biomass on a regional basis,[14] and fuel cells do not necessarily have to run on hydrogen. Other options are:

- Biomass-derived alcohol (ethanol). This does not have to come from distant tropical sugar cane plantations; alcohol can also be produced from sorghum, or from the lignin component in wood. It has been calculated that 1 tonne of dry wood mass can yield 387 litres of ethanol.[15]
- Gasified biomass. Gasification plants can cope with all forms of biomass, so there would even be no need to separate or otherwise prepare the material beforehand, making it possible to utilize the entire plant.[16]
- Methanol, which can be synthesized from plant-derived carbon and hydrogen.[17]
- Biological petrol subsitutes, which can be synthesized from gasified biomass and hydrogen, which in turn can be produced using local wind power plants.[18]
- Biogas, which results from anaerobic fermentation of organic waste.

Clearly, there is no particular need to put all our eggs in the fuel cell basket; internal combustion engines can be made to run on vegetable oil instead of diesel, gasified biomass rather than natural gas, or on bioethanol or biomethanol. An extensive discussion of all these, and other, various options is not possible here, but there is a wealth of literature and calculations available, all of which can already draw on practical examples.[19]

Synergistic applications, cross-substitution and all-load micro-power plants

The importance of power storage lies in the opportunity for complete, efficient and cost-effective self-sufficiency in electricity, heating and even fuel. Generators for domestic power production have been around for as long as there have been motors. Up till now, electricity production from such generators has only been a viable proposition in comparison to mains electricity if combined with heat production. However, the mutual interdependency of heat and electricity production in a cogeneration plant and the consequent problem of over- or underproduction has still made a grid connection necessary, so that surplus electricity could be sold to the grid (usually below cost) and additional electricity purchased from the grid to make up any shortfall. The ability to store electricity solves the problem of over- or underproduction for operators of CHP plants, and consequently their dependency on the grid. Any surplus can be stored, be it heat or electricity, because heat can also be converted into electricity. All stored electricity can be used, whether to meet the electricity needs of the building, to run a car or to supply heat.

Decentralized energy storage is the link that enables electricity generation from renewable sources to be combined with motor technology in an optimal manner. The motors and drives industry* is potentially a staunch ally of renewable energy – it need only refocus on the production of motors that can be run directly from renewable sources or from stored energy. The outcome would not be one single type of motor, but a wide variety of types, depending on the particular energy source and energy generation scheme.

If autonomous energy supplies are to achieve broad appeal and develop a momentum of their own – to become a 'solar steam-engine' – then the number of interlocking technologies must be kept small. With a domestic-level cogeneration plant in combination with a power storage system, residents could become independent of external electricity supplies and produce their own motor fuel. In this case, all they would need would be a supply of fuel for the cogeneration plant, the lynch-pin in this scheme.

If they have a fuel cell car, the on-board fuel cell could be combined with a power storage system to generate electricity while the car stands idle in the garage, with sufficient reserves

---

* Manufacturers of electric motors and drives for industry (a drive being essentially an inverter-based regulator for an electric motor).

of power being laid in to cover times when the car is in use. The power generation plant then comes with the car. If a PV and electrolysis plant is added in, all or some of the fuel for the car can be domestically produced.

These are only a few examples among many of what could be achieved in future. Buildings whose heating, electricity and fuel needs are met solely from the sun, independently of the grid, are also conceivable.

Buildings as energy collectors and generators

Buildings do not just consume energy: in future they must also be seen as systems for collecting and harnessing solar energy. Among other things, conservationally minded architects or planners must give thought to:

- optimal solar gain, by allowing as much daylight as possible into the house. Besides windows, this might mean novel light storage systems;
- exploitation of ambient heat as a secondary heat source; siting the building so as to make maximum use of the sunlight and avoid shading; windows to open up the sunny side of the building and insulation on the other sides; optimal locations for solar collectors and PV arrays;
- natural ventilation from prevailing winds on the site;
- selecting building materials according to their insulating properties and the energy cost of their manufacture – ie, determining where concrete and aluminium can be replaced by wood, rammed earth or steel; and
- internal air circulation, which can also be used for seasonal cooling or heating.

In the end it comes down to what the building engineer Klaus Daniels calls 'building to the climate': 'What distinguishes intelligently planned and operated buildings above all is their ability meet their occupants' needs directly from the environment without hi-tech equipment, through natural lighting, natural ventilation, means of controlling the total amount of energy or daylight entering the building, and so on.'[20]

Table 6.3 *Energy regulation strategies in biological systems, compared with existing and potential architectural applications (selected examples)*

Biological energy regulation strategies	Existing and potential architectural applications
Control of heat loss through capillary dilation/contraction	Adjustable exterior surfaces
Contraflow and heat-exchange (eg whale flippers, seabird legs, gills)	Heat exchangers in ventilation shafts and wastewater pipes
Insulation through zoning and restricted blood circulation in cold conditions (eg in arctic animals and insect nests)	Frequently used in traditional architecture (eg farmhouses in Lower Saxony and South Germany, conservatories)
Pigment alterations to make best use of solar heat	Regulation of light absorption and reflection using blinds or glass roofing in greenhouses
Expansion and contraction of insulating layers (by ruffling up feathers, winter and summer coats)	Adjustable insulation, summer and winter dwellings in traditional architecture, winter windows etc
Light absorption for heat gain (translucent skin or hair, eg alpine plants, polar bear)	Modern transparent insulating materials, glass façades
Maximizing heat collection through expansion of the collecting surfaces (by opening wings or exposing flanks to the sun)	Adjustable façades, porches
Evaporative cooling (sweat, panting, transpiration in plants)	Cladding which conducts or radiates away excess heat
Shading to promote convection (eg cactus ribbing)	Suitable shade-producing exterior structures
Light filtering and distribution (eg South African window plants)	Glazing with different colouring and material properties, grills, ribbing
Piped-in light (light channelling in plant shoots)	Daylighting systems, light pipes
Ventilation (termite mounds, prairie dog warrens)	Chimneys, wind towers (Arabia, Iran)
Centrally regulated heat exchange (skin, behaviour)	Computer-controlled solar buildings with an optimal combination of energy-management measures

Source: Helmut Tributsch, *Wohnen mit der Sonne* (Living with the Sun)

The European Charter for Solar Energy in Architecture and Urban Planning, drawn up under the overall control of the Munich-based architect Thomas Herzog in 1996, lists all these design options in detail.[21] Helmut Tributsch of the Hahn-Meißner Institute in Berlin has attempted to illustrate the energy-saving potential of intelligent materials using examples of energy regulation in biological systems.[22] He calls upon architects to learn from the self-regulating energy systems of the natural world, and to copy them in their choice of material and form. This suggests scope for countless new multifunctional materials (see Table 6.3). The vast range and variety of possibilities that energy-autonomous buildings offer their occupants is clear. It is also clear that we are only just beginning to realize them.

It is not simply a case of bolting on technology, but rather a new form of architecture. In the words of Sir Norman Foster, describing prominent examples: 'There is an extraordinary elegance that emerges from a true response to climate and place.'[23] Building components acquire multiple roles: not just roofs, walls, windows, they also collect, store and exchange energy. The result is a renewed diversity of architecture. As the greatest part of fossil energy consumption takes place in buildings, solar construction techniques become part of the responsibility of society at large to maintain the environment, and part of the responsibility of business at large to reduce the need for imported energy. If buildings increasingly can become autonomous in energy, and the direct need for external supplies of electricity and heat can be reduced towards zero without compromising on lighting and heating, then the built-in solar plant begins to take on the role of power stations, able to sell surplus energy. Energy costs become energy gains.

The technical know-how to revolutionize energy supplies

The technological options described here are the kindling for the energy supply revolution to come.

1   The various available sources of energy and autonomous generation and storage technologies make it possible to deploy energy and technology for multiple applications. In particular:
    • energy sources can be mutually interchangeable, ie electricity can be replaced with heat or fuel, heat with

electricity and fuel, fuel with electricity and heat, thus allowing energy to be used more flexibly and more efficiently;

- individual technologies – heating, cogeneration and storage plant, PV and collector arrays, equipment and vehicles – can be applied to a variety of functions, thus allowing more productive deployment of technology.

2   Despite rising electricity consumption, the demand for grid electricity will fall if ever more devices produce and store their own electricity. The same applies to distributed heat, due to the scope for solar-optimized construction techniques.

3   Combining electricity storage with motor-driven generators makes it possible to construct local micropower plants capable of serving all loads, rather like a car with an automatic gear-box which shifts gear from neutral up to fifth as required. This renders the load management regimes of the grid-based electricity industry – whereby electricity is supplied from multiple power stations with varying output profiles – superfluous. The costs of maintaining the grid are replaced by the costs of electricity storage and reserve capacity, which will deliver true cost transparency for the first time ever in the energy industry. Once the advantages begin to be realized, the existing electricity industry's days are numbered. Pylons will cease to 'march' across the landscape.

4   The familiar pattern of centrally managed supplies of electricity, fuel, heat and process energy will be replaced by a comprehensive decentralized supply.

5   The essential prerequisite for all this is the deployment of renewable energy, because only from renewable sources can energy be supplied overwhelmingly without long supply chains. Only then can decentralized technologies lead to a truly decentralized energy system. The quickest and easiest way to meet electricity and heating needs will be through renewables. Fuel can also be supplied from local sources – that is, if the choice of vehicle motor technology does not lead to fuels being replaced by electricity.

It is not enough merely to see the possibilities: we must put them into practice on the ground, be determined to learn how they can shape our lives, and, module by module, we must build on that potential. There will be no shortage of individual and new community strategies for energy supply at local and regional levels. Every new technological and conceptual advance strengthens their economic and cultural appeal, and increases their power to throw off the resource supply chains of the conventional energy system.

The solar technology revolution and the solar information society

At first glance, local or individual energy autarky would seem to require a complex structure, and to be impractical and thus unrealistic to achieve on a broad scale. This impression arises, however, because the public has not yet had enough time to become aware of the slew of new technologies and adjust their habits to suit the new ways of working. On closer examination, energy autarky is no more complex and thus no less convenient than the current structure of energy supply.

Once an autonomous decentralized energy supply is in place, it will demand less of the consumer in terms of costs and personal initiative than do conventional energy supplies today. Obtaining energy is not just a matter of putting the plug in the wall and turning on the switch: the electricity bills must be paid and checked; various gadgets and appliances need spare batteries or time must be allowed for recharging; the boiler needs maintaining, gas or oil must be ordered for the next heating season; the car must be refilled regularly, and there are oil changes, maintenance and emissions check to think about. In future, there will also be the need to compare prices to find the cheapest electricity supplier, as is increasingly happening in the telecoms sector. By comparison, the difficulties with solar energy technologies are the current paucity of information and advice, and the fragmented and uncoordinated way that the technologies have so far been introduced. Coordination could be made considerably easier through the use of IT and associated services, which would save the energy consumer having to source all their energy services individually.

The full impact of new technological solutions is only felt when their promoters go beyond the familiar structures of technology use. If those opposed cannot halt the development in its tracks, the result is a structural revolution that former critics then try to use to their advantage. This was the pattern for the growth of IT, with both positive and negative social and environmental consequences. The negative consequences arise because these distributed technologies have not been truly independent of centralized infrastructure, which has allowed the established corporations to regain control.

In fact, the way IT and technologies for generating and harnessing solar energy can complement each other makes them ideal partners. Solar power can liberate IT from fixed energy supplies, allowing it to be deployed in an even more mobile and independent way, whereas IT can make solar devices smarter. Microelectronics allows different devices to communicate with each other, and programs can be developed to integrate and control them. Autonomous solar energy supplies and multifunctional systems can be controlled by remote data link, and simulations can be run, analysed, modelled, halted and maintained.[24] The transition from a national energy grid to independent mini-grids thus becomes considerably smoother. Today's PCs can perform tasks that, until the 1970s, only mainframes could manage. Likewise, small solar energy systems could in future take over from the industrial-scale technology of nuclear and fossil energy supplies.

More than anything else, what characterizes the economic and societal modern age is the idea of the 'information society'. The concept arose with the advent of IT, and the pundits have been singing its praises ever since. One such information society evangelist is John Naisbitt, who sees in it the 'machine of individualism' which will radically reshape all economic and political structures: 'the deployment of power is shifting from the state to the individual. From vertical to horizontal. From hierarchy to networking.'[25] The larger the world economy grows, the more powerful the 'small players' become, while the 'big players' decline in importance. This, allegedly, is the 'global paradox'. With the 'mixed technologies' of the telephone-cum-television-cum-computer hybrids, anybody can communicate with anybody from anywhere. The 'personal telecomputer', he claims, will bring 'wireless productivity' through technologies that become ever cheaper, lighter, small and more mobile. Ultimately, 'the most efficient and most effective economic unit will be the individual', who will work within a network that itself is part of a global network in which no one company and no one country can be a successful player in the 'global game'. Hence the urgent need for global players to form strategic alliances.

This, however, is the great illusion of the information society. All the technical details are correct: fast communication from anybody to anybody, the lack of geographical ties. But power is not moving sideways, but upwards; and the centres of power are not shrinking, but becoming larger and more hierarchical than before. This contradiction between the centripetal momentum latent in the technology and the centrifugal agglomeration of power that is actually taking place stems from the fact that the centres of power can make better and faster use of the information available, because their greater organizational and financial muscle offers better opportunities to turn words into deeds. Where a technological network exists, someone has to run it. The 13 largest internet service providers in the world are all US-owned. In the interest of creating a European counterweight, the activities of electricity companies are accorded special treatment by politicians. The 'invisible hands of the networks' writes Philippe Quéau in *Le Monde Diplomatique*, 'are weaving their own uniform web. The functional logic of the network favours mergers and synergy effects – in the language of the market: collusion, oligopoly and monopoly.'[26] Competition will continue to bring prices down, until the trend turns towards mergers and industrial concentration. Information may still flow freely, but already there are those with privileged access, and as monitor and television screen come together to form a single gateway for information retrieval and receiving TV programmes, the rich array of information providers contrasts with one-sided presentation through the mass medium of television. The network, without which information technology could not function, creates new dependencies that call the new autonomy it offers into question and reduce its meaning. Networks can liberate, but they also always represent a restriction on or danger to individual freedom. Networks can be ties that bind.

This is what sets localized solar power apart from the applications of information technology. No-one controls the solar 'broadcaster'. Sun and wind do not need wires and transmitters to broadcast their energy. When harnessed through autonomous installations, they make networks dispensable. Moreover, this dispensability erodes the foundations of the ever-expanding networks of corporate power that the energy supply chains made possible.

Independent solar technology can realize what IT promised: the creeping decline of global corporate power, the whittling down of business hierarchies. The technology is still in its infancy, comparable perhaps with the automobiles of the 1920s. Work on PV is aiming at producing materials that can absorb far more sunlight and which can reduce material inputs a hundredfold. Cost reductions will result. The solar cells of the future will be made of flexible material capable of turning even the smallest quantities of light into electricity. Photoactive materials ('wet solar cells') may be developed using photochemical processes; photolytic separation of water; miniaturized electrolysis plants and miniature fuel cells; light concentrators; ultra-light light converters; high-performance thin insulating materials; fluvial hydropower; mega- and mini-wind turbines which can make use even of light winds; improved biomass gasification plants; high-performance small Stirling engines...[27] – these and other developments are in the pipeline, on top of which comes the potential for cost-reduction through mass production and improved tooling. The solar–technological revolution has only just begun. The driving force is the practical applications – the introduction of technology and its uptake by society.

Trains will be pulled by fuel cell locomotives, eliminating the need for overhead cables and making railways cheaper to run. Carriages will have solar modules built into the roof. Already, there are refrigerated lorries whose roofs are covered with PV panels. There will be airships whose entire external skin is one big photovoltaic array, which will provide much of the power they need. Freighters will produce their fuel on board, through wind power and electrolysis, and passenger ships will also use biogas plants to process their organic waste. What is needed is imagination and – more importantly – new priorities for science and technology, for architecture and for energy supplies themselves, for companies and governments. What is required is a vision for an energy supply that does not rely on a specialized energy industry.

7

The untapped wealth of solar resources

THE BIG PROBLEM with current environmental policy and environmental management procedures in businesses is that individual problems are tackled in isolation. The result is an unmanageable catalogue of single-issue demands and measures. Even by the mid-1980s, the Chemical Abstract Service had registered 8 million chemicals, mostly synthetic,[1] and well over a million must have been added since. Several hundred thousand of these registered compounds are in active use. Even if only 1 per cent of products cause environmental difficulties, this makes proper environmental safeguards next to impossible to achieve, regardless of whether the instruments used are laws, regulations or voluntary agreements in industry. Even enhanced resource productivity is only of limited help, as productivity gains primarily reduce waste, without affecting the number of products produced or the raw materials used. Fossil resources require a thicket of specific regulations which no-one can hope to enforce and which smother both business and personal lives in red tape.

It is therefore no coincidence that public sympathy for environmental protection and environmental policy is waning, despite general recognition of the dangers. Careful consideration of every aspect of each environmental problem becomes tiresome and stultifying. The situation is also often worsened by failure to distinguish between truly serious and more trivial problems. Besides which, many people do not see why they should take an environmentally responsible attitude towards small problems while global problems go unchecked, often as not made worse by governmental action. As people lose patience with the vast range of demands made upon them, the cause of environmental protection is undeniably losing ground.

We need to radically rethink the relationship between industry and the environment, and harnessing the plant kingdom to provide industrial raw materials is the ideal way to do it. Changing the energy and resource base goes to the root of the whole problem of limited resources and the damage caused by their use. Historically, environmental concern in the energy sector meant energy conservation and energy efficiency. Now, as the technology improves, the focus must shift towards renewable energy. Likewise, in the case of industrial raw materials, the priority now must be to take a long, hard look at the replacement of limited reserves of raw materials with renewable solar (ie, biological) resources. Tapping the unlimited potential of solar materials will make it possible to move from rearguard criticism of environmentally dangerous activities to practical support for environmentally neutral industrial processes and products. The transition from fossil to solar resources is just as important as that from fossil to solar energy, and what is more, it is just as practicable. In many cases, it may even be easier to achieve.

Following a century in which plants and vegetation have been pushed to the margins of economic and cultural life, the last quarter-century has seen a revival of interest in this biological treasure chest. After long years in which urban greenery has had to make way for asphalt and concrete, and in which avenues of trees were even thought a danger to traffic, plants are in many places now making a comeback, both for aesthetic reasons and as a tool for improving the local microclimate. Deforestation and its consequences has become one of the global issues of the day. Reforestation initiatives are appearing, albeit they cannot keep pace with the continuing destruction of forests across the globe. By no means is it just the tropical rainforests that are disappearing: large tracts of North American and Siberian woodland are also under threat. There is now increasing recognition of the value, and even more, the potential value of biodiversity. Not least among the new botanical prospectors is the chemical industry – not that chemicals companies have made much effort to put a stop to the continuing destruction of biodiversity that results from climate changes wrought by fossil fuel consumption.

As fossil fuel reserves near exhaustion, the chemical industry must harness biological resources if it is to survive. Nevertheless, with few exceptions, chemicals companies are jeopardising their future existence by remaining firmly wedded to their fossil fuel resource base. 'As long as there is a choice between fossil and regenerable resources,' states a BASF representative in defence of this strategy, 'the regenerable resource must be competitively priced on the world market and in adequate supply.'[2] This attitude illustrates the barrier that industrial myopia and structural conservatism represent to the industrial potential of solar resources.

The transition to a solar resource base would not only allow the 'poisoning of the planet' (Karl-Otto Henseling)[3] by chemical production methods and chemical products to be largely avoided. Although evolution has also brought forth all manner of highly dangerous toxins, nevertheless, the natural world in its entirety is a well-constructed, sustainable mechanism, because natural poisons are biodegradable and optimized to perform highly specific functions. Chemicals companies are already trying to learn from nature by combing through rainforests, for example, in search of plants that have evolved their own chemical defences against insects. Solar resources also provide opportunities for productivity gains that fossil raw materials could never deliver. There are already convincing examples from the world of 'natural chemistry' of how photosynthetic processes produce countless compounds with molecular structures that synthetic chemistry can replicate only through laborious, highly toxic and highly complex procedures.

Many people argue the case for renewable energy on the basis that burning fossil fossils is a waste of an indispensable raw material for the manufacture of synthetic goods. However, this well-intended reasoning seriously underestimates the damage done to the global ecosystem by chemical production processes, which consist overwhelmingly in the conversion of fossil hydrocarbons – ie, crude oil, natural gas and coal – into chemical feedstocks. It also underestimates the potential for solar materials as a comprehensive and, in many respects, superior alternative to synthetic petrochemicals.

Biomass gasification produces a gas which is just as suitable for syntheses as the natural gas which is currently used – with the important difference that biomass-derived gas is almost devoid of sulphur. With biogas, the chemicals industry could retain existing synthesis pathways while simultaneously reducing their environmental impact. As plants are composed of hydrocarbon compounds, logically, anything produced from fossil hydrocarbons could also be produced from plants. Yet biological raw materials carry very different implications for processes and products than their fossil counterparts. In many cases, as with the substitution of renewable energy for nuclear and fossil fuel power, the shift towards solar materials will overturn existing procedures and reshape business relationships. Moreover, there is considerable scope within the spectrum of solar materials for reducing reliance on metals.

Whether and how these opportunities are seized is again a structural issue. Isolated replacements of conventional by solar materials in the production of particular intermediate or final products are of course sensible steps forward in their own terms. But the full scope for productivity gains from solar materials by comparison with fossil resources really comes from the very different ways the two are produced. This is what makes it clear that it is the *choice of resource base* that matters, not the actual quantities used. It also shows that the current slew of biotechnology companies have grabbed completely the wrong end of the stick.

The higher productivity of biological materials

The long road from extraction of oil, coal or gas, through refining and complex industrial processes to finished chemical feedstocks is matched in the case of solar resources by planting, harvesting, storage, cleaning, drying, separation and transport. In many cases, transport costs can be held to a minimum by growing the materials near the production plant. The only disadvantage as far as productivity is concerned is the considerably greater need for personnel, although this also represents a significant social benefit. Giuliano Grassi has calculated that, for each terawatt hour of energy, natural gas requires 250 employees, crude oil 260, coal 270 and nuclear power 70, whereas producing solid fuels from crops requires 1145 employees and 1000 employees in the case of woodlands.[4] The greater labour input probably also applies to the use of solar materials in the manufacture of chemical products in place of the equivalent fossil feedstocks.

But against this alleged disadvantage, which takes into account the resource costs but not the costs associated with chemical industrial processes, should be set a raft of economic advantages. The industrial processing required for solar materials is much less extensive than for fossil hydrocarbons. The great diversity of natural chemicals means that, in many cases, they can be used 'as is', their chemical structure needing little alteration. By comparison, crude oil in its raw state is not even halfway usable as a chemical feedstock.

Hermann Fischer, a manufacturer of natural paints and dyes whom *Capital* magazine accorded the title Eco-Manager of the Year in 1992, illustrates this using the example of polyurethane, the base material for varnishes, adhesives, foams and artificial fibres.[5] Crude oil requires substantial chemical engineering to turn it into usable feedstocks. The process involves high temperatures and chemically active surfaces, such as those based on the heavy metals. First of all, the sulphur content has to be removed by catalytic hydrogenation. Unwanted aromatic compounds are removed, and the purified hydrocarbons are then transformed into the desired molecular structure in a washing and reforming process at temperatures of between 500 and 1000 degrees Celsius. Component molecules are added to or removed from the basic structure as required. Only after this molecular disassembly has taken place can the hydrocarbons be turned into desirable precursor chemicals. Converting these precursors into active reagents results in highly toxic by-products. For example, one by-product of the process used to manufacture chlorinated hydrocarbons, the base material for the manufacture of pesticides, preservatives and wood treatment agents, is extremely toxic phosgene gas, which can be (and has been) used as a chemical weapon.

It is a combination of these reagents that produces the actual polyurethane, which is itself a feedstock for numerous other products. The component molecules are released when the final product is consumed or as it breaks down, gradually evaporating, peeling, ablating or being absorbed by living tissue. Even if the final product does not itself contain poisonous substances, this should not distract from the fact that highly toxic chemicals are employed in manufacturing the intermediate products, and that industrial chemistry is associated with considerable health risks and produces highly toxic waste. Where the final product ends up as waste, the very features that endow it with a long shelf life and which make it easy to use also cause it to persist in the environment. Reverse engineering back to mineral substances is energy-intensive, complicated and rarely performed for cost reasons.

For example, for every 100 kg of benzopurin 4B dye produced, there are 768 kg of waste and by-products. These figures relate only to the last stages of the manufacturing process – the precursor substances, from crude oil through to the production of naphthalene, aniline and toluol during refining are not included.

As Figure 7.1 makes clear, industrial hydrocarbon chemistry is an open loop (Fischer calls it the 'petrochemical snake'), and therefore cannot offer fundamental economic advantages over its biological counterpart. Figure 7.2 underlines this point by comparing fossil and biological production pathways. Given its obvious drawbacks, it is perverse that the EU framework agreement of 1992 gives preferential treatment to the established structures of the chemical industry as a matter of course, by making 'crude oil consumption by the oil-processing industry' – ie, the chemical industry – tax-exempt. This amounts to environmentally detrimental and market-distorting subsidies of the order of billions – legitimized, once again, by the presumption that there is 'no alternative'.

The reason given for preferring petrochemical over solar materials is the latter's higher price. On a closer examination, however, this cannot be the determining factor – at the very least, it would carry little or no weight if the tax-exempt status of crude oil were revoked. Around 300 precursor chemicals are produced from oil, coal and gas. The waste disposal costs of these alone would be enough to tip the balance in favour of

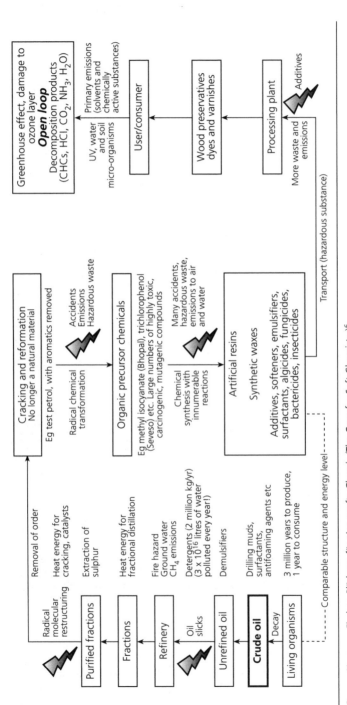

Figure 7.1 *The 'petrochemical snake'*

Source: Hermann Fischer, *Plädoyer für eine sanfte Chemie* (The Case for Soft Chemistry)[6]

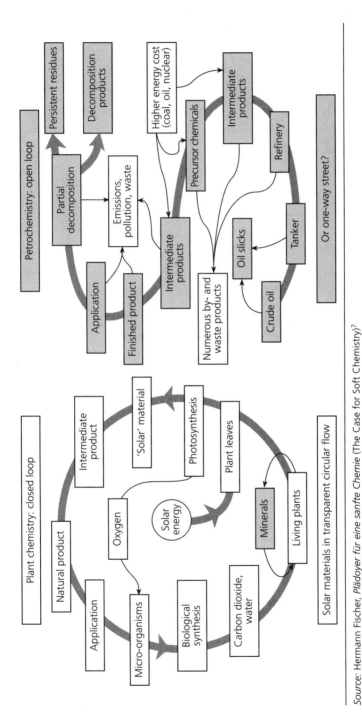

Figure 7.2 *Comparison of solar and petrochemical resources*

Source: Hermann Fischer, *Plädoyer für eine sanfte Chemie* (The Case for Soft Chemistry)[7]

Table 7.1 *Comparison of market prices for fossil and regenerable resources*

	Fossil	€/t	($/t)	Regenerable	€/t	($/t)
Raw materials	Crude oil	112	(99)	Maize	77	(68)
	Natural gas	171	(152)	Wheat	99	(88)
	Naphtha	137	(121)	Soya	214	(190)
Precursor chemicals	Benzole	248	(220)	Rape seed oil	445	(395)
	Ethylene	332	(295)	Palm oil	317	(282)
	Propylene	245	(218)	Molasses	133	(118)
	Methanol	97	(86)	Sugar	399	(355)
	Ammonia	91	(80)	Starch	274	(243)
Intermediate products	Ethylene oxide	511	(454)	Sorbitol	640	(569)
	Propylene oxide	921	(818)	Glycerine	1023	(909)
	Acrylic acid	1100	(977)	Citric acid	1023	(909)

Source: KATALYSE Institut: *Leitfaden Nachwachsende Rohstoffe* (Handbook of Renewable Resources)[8]

solar resources, and the outcome becomes even more favourable once the damage to the health of the petrochemicals workforce is brought into the equation.

The *Leitfaden Nachwachsende Rohstoffe* (Handbook of Renewable Resources) compares the market prices for fossil and solar resources. A summary is presented in Table 7.1.

The table distinguishes between raw materials, precursors and intermediate products, and compares fossil and solar substances used to manufacture comparable products. The table also shows that, whereas fossil raw materials require complicated processing to produce the required precursor chemicals, in the case of solar resources, the precursor is the natural state of the raw material – or only requires simple processing to extract the sugar, molasses or starch. Two of the solar raw materials listed, and even more of the precursors, are already cheaper than their fossil equivalents; the price differences for the intermediate products are small. The slightly higher prices for solar intermediate products cannot be because it is fundamentally more expensive to process solar raw materials into precursor chemicals; the opposite is the case. The real reason for the higher costs is to do with quantity. As more care can and must be taken when selecting particular solar precursor or intermediate products, there is less scope for economies of scale than with fossil resources. In any case, industrial processing of solar materials is still in the early stages.

There are already many practical examples of the multifarious productivity and environmental benefits that can flow from the industrial exploitation of solar materials, as Hermann Fischer documents in his book *Plädoyer für eine sanfte Chemie* (The Case for Soft Chemistry). Fischer uses 34 criteria to perform a systematic comparison of fossil and solar resources as employed in the production of paints, covering source raw material through to product use and disposal. Only on one criterion, ease of use, do synthetic paints outperform natural ones, and this is primarily the result of a century of industrial research and development. Natural paints could easily catch up if industry and academia were to devote more time and effort to the issue. On only one further criterion – shelf-life – do both perform equally; on all the other 32 criteria, natural paints have a significant edge (see Table 7.2).

The economic advantages of solar resources are also documented by a large number of examples in many other areas. For instance, the car manufacturer Daimler has concluded, after experimenting with plastic reinforced by natural fibres, that natural fibres require far less energy to process than glass fibre. Everything can be recycled without significantly compromising on the technical specifications – of particular relevance to the automobile business, given the impending legal obligation on car manufacturers to take responsibility for old cars. A weight reduction of 10–30 per cent favours energy-efficient light construction techniques. Natural fibres are also beneficial to workplace health, as they cause less skin and lung irritation. Lower prices and reduced processing time also reduce the cost of manufacturing the respective components by 10–30 per cent.[9] Industry also reports numerous advantages from its experience with lubricants derived from vegetable oil: reduced evaporation, less wear and tear on motors or machinery, longer lifetime, no microbial decomposition, no waste disposal costs, reduced materials maintenance due to better sealant properties, no risk to water quality, more favourable temperature/viscosity relationship, no need for continual supervision. These biolubricants cost between two and five

Table 7.2 *Comparative evaluation of products manufactured from fossil and solar raw materials*

Criterion	Synthetic colours	Mark	Natural colours	Mark
	Colours from crude oil or colours from plants? (Comparative environmental impact assessment of the main components in paints: adhesives, solvents, organic pigments, additives)			
a) Raw material				
Origin	Crude oil	− −	Plants	++
Renewable source?	No	− −	Yes	++
Availability	Very limited	− −	Unlimited	++
Global distribution	Very uneven	− −	Near universal	++
Producers	Corporate monopolies	− −	Regional producers	++
Toxicology	Highly toxic	− −	Mildly or non-toxic	+
Environmental impact	Highly damaging	− −	Harmless	++
b) Synthesis				
Basic principle	Industrial synthesis	− −	Photosynthesis	++
Where synthesized	Reaction chamber	− −	Within plants	++
How organized	Centrally	− −	Locally	++
Energy requirements	Very high	− −	Low	++
Energy source	Oil, coal, nuclear	− −	Direct solar energy	++
Chemical structures	Highly artificial	− −	Natural structures	++
Process control	Costly	−	Self-regulating	++
Risk of accident	Very high	− −	None	++
Effect of accidents	Sometimes disastrous	− −	None	++
Security requirements	Very high	− −	Very low	++
Fault tolerance	Very low	− −	Very high	++
Emissions	High	− −	None	++
Quantity of waste	Very high	− −	None	++
Type of waste	Highly toxic in parts	− −	(Oxygen)	++
Social costs	Very high	− −	Low	++
c) Product use				
Chemical pollutants	Can be considerable	− −	None	++
Effect of inhalation	Mostly harmful	−	Mostly harmless	+
Odour (solvents and ancillary chemicals)	Overpoweringly artificial	− −	Pleasantly natural	++
Colour aesthetics	Loud, glaring	− −	Lively, harmonious	++
Tactile quality	Unpleasant, smooth	− −	Pleasant to the touch	++
Electrostatic properties	Collects a high charge	−	Charge usually low	+
Effect on indoor climate	Often negative	−	Highly positive	++
Ease of use	Good to very good	++	Good	+
Durability	Good	+	Good	+
d) Disposal				
Biodegradability	Sometimes very poor	−	Complete	++
Rate of decomposition	Sometimes very slow	−	Very fast	++
Atmospheric impact	Greenhouse effect	− −	None (circular flow)	++

Source: Hermann Fischer, *Plädoyer für eine sanfte Chemie* (The Case for Soft Chemistry)[10]

times as much as standard fossil hydrocarbon lubricants, but scarcely more than fossil hydrocarbon lubricants of equivalent quality. Even with the current state of knowledge, 90 per cent of all lubricants could be replaced with ones derived from vegetable oil.[11] Even today, biolubricants have an economic advantage once the qualitative advantages have been taken into account.

Only in the realm of medicine have plant materials made significant inroads. Plant-derived remedies, produced from a wide variety of medicinal herbs, already make up 30 per cent of the market for medicines in Germany.[12] Henkel is one of the few large chemicals companies which also uses solar materials in its other products, from natural-fibre-reinforced plastic rather than fibreglass through surface chemistry to adhesives, cleaning agents and cosmetics. For some products, cleaning agents in particular, the entire product line is now manufactured from solar sources.[13] Already, one can point to a panoply of further examples, involving different solar resources in each case: chitin, extracted from insect exoskeletons or from fungi, is used to make the polymer chitoson, which has applications, among others, as a preservative, in catalytic converters and as packaging; new materials derived from pea starch; the use of starchy potatoes, whose juice contains nitrates and potash, in composite materials, space-filling packaging and cushioning material or plastic film; the use of plant fibres and cellulose in insulation and foams; packaging materials made from maize; the many and varied uses of hemp or flax in textiles, building materials, medicines or paper; high-performance fibres made from banana stems. In all cases, the products are biodegradable and can be recycled naturally.

Native plants already contain usable materials which it would require highly toxic and complex procedures to produce from fossil resources. As Fischer puts it, plants:

> never [form] a single substance or even a single class of materials. Even the simple conifers simultaneously synthesize cellulose, lignin, dyes, tannins, chlorophyll, vegetable waxes in the cuticle of its needles, essential oils, resins, turpentines, oil- and protein-rich cones and thousands of other easily used

and sought-after natural materials. Every single species of most plant genera offers almost the entire range of plant products. It therefore makes sense not to harvest a plant for the sake of just one material, but as far as possible, to use every part of it.[14]

Solar resources offer obvious advantages for industrial processing. To exploit them, however, would necessitate a departure from the industrial monoculture of the chemical industry. Solar resources, therefore, represent an opportunity for small and medium-sized enterprises to establish themselves on the market for chemical products.

The chemical industry in Germany consumes 13 million tonnes of crude oil, 2.7 million tonnes of natural gas, 1.5 million tonnes of coal and 1.8 million tonnes of vegetable matter annually. The latter is primarily used to produce 450,000 tonnes of starch, 250,000 tonnes of cellulose and 900,000 tonnes of vegetable oils and fats.[15] The enormous potential of solar resources has hitherto not been fully exploited. There are a wide variety of reasons for this. Manufacturers of biochemical products are mostly small, and are forced, through lack of large marketing arms, to rely on specialist markets. Chemistry research is overwhelmingly funded by the petrochemicals industry. And the knee-jerk prejudices and uninformed nature of the environmental movement, which specializes in conserving nature without utilizing it, have held campaigners back from forcing a fundamental rethink in the chemicals industry.

Replacing fossil with solar resources

According to Römpps Chemical Lexicon, the global chemical industry consumes around 900 million tonnes of fossil raw material every year. By comparison, the biosphere produces 1.7 trillion tonnes a year on the land surface alone, almost 2000 times greater than the entire annual demand for petrochemical products.[16] At the fourth international conference on Solar Energy Storage and Applied Photochemistry in January 1997 in Cairo, it was calculated that the chemical industry's total annual output of materials is only 0.02 per cent of the total annual output from nature.[17]

The UNEP's Global Biodiversity Assessment cites 10 to 100 million plant species.[18] In his book *Nutzpflanzenkunde* (Crop Science), Wolfgang Franke writes of 400,000 known species, of which 20,000, or 5 per cent, are used as foodstuffs, medicines, for pleasure or as raw materials. However, only 5000 species are cultivated, and only 660 are major agricultural crops.[19] Few people have recognized, and then only in part, the number of uses to which a plant can be put. Even isolated examples expose a fascinating panorama of possibilities. In all likelihood, therefore, the variety of products which can be made from the 'rediscovered' hemp plant, as collated in a much-respected book,[20] was only the tip of the iceberg. The book listed only historically documented uses. How many more await discovery will only be known when materials scientists finally devote real research time to the subject. Several hundred new products might quickly result. The same probably goes for almost all the 400,000 known and innumerable unknown plant species.

It may be impossible to put a figure on the full spectrum of possible materials. It will greatly exceed the number of registered chemicals. Estimates to the effect that science has uncovered considerably less than 1 per cent of what is actually possible are not overly conservative. Part of this knowledge is which plants produce the greatest and qualitatively most valuable yield for which purpose under what conditions. Yet instead of seeing this potential as an opportunity for a comprehensive greening of the chemical industry, the world is being held by the fossil energy and materials industry in a state of economic monoculture which stands in ever more dramatic contrast to the growing diversity of opportunity. It would doubtless be possible, compound for compound, to find photochemical replacements for the entire spectrum of petrochemical products.

But what about the other group of non-renewable resources, the mineral ores? In its publication *Mineralische Rohstoffe. Bausteine für die Wirtschaft* (Mineral Ores: Components for Industry), the German Federal Institute for Geological Sciences and Raw

Materials lists substitute materials for 12 important metals, of which seven have already been partially replaced. Casings, pipes, oil pipelines, filters, household appliances, windows and doors, sewers, gutters and packaging are already being made from plastic rather than copper, aluminium, brass, steel, lead, zinc or tin.[21] It follows that they can also be made from solar materials. The list could be extended ad infinitum: from car bodies to aeroplane fuselages and wings, from ship hulls to tanks and cables. One of the greatest opportunities is the replacement of aluminium by wood as a construction material – it is no coincidence that wood-frame buildings are experiencing a renaissance.

All these examples are no more than snapshots. Solar materials are poised for a new beginning and an all but unlimited new realm of scientific and technical possibilities. As this development gets in train, further opportunities for replacing metals will open up. Even if not all metals can be replaced (for example, those which are good conductors or highly refractory), solar materials present a real opportunity for a drastic reduction in the consumption of conventional raw materials.

Solar materials: from agricultural monocultures to polycultures

The chemical industry's argument that the 'costs are too high' is not the only objection raised to solar materials (and by extension to the replacement of fossil resources). There are a further two stock arguments drawn from the conservation debate, and these also crop up time and again in connection with the use of biomass as an energy source.

I   The first argument is that solar resources compete with food production for land, hence there are 'ethical reasons' for sticking with fossil petrochemistry, because otherwise there would not be enough arable land to produce the necessary quantities of food. This reasoning, however, is not tenable, as the data on worldwide arable land area in Chapter 2 indicate. Natural photosynthetic production is entirely capable of replacing the third of annual oil output consumed by the chemicals industry, and quite possibly more besides.

2 Secondly, critics cite the danger of over-farming the land and of agricultural monocultures. Yet it is doubtful that these tendencies would automatically be exacerbated by the use of solar resources as an energy source and as raw materials. From the earlier discussion on solar resources as an energy source, the conclusion has already been drawn that this is not the case, provided that such energy crops are harnessed in the technologically and economically optimum manner.

Studies such as the one by David O Hall and Frank Rosillo-Calle show that the danger of soil erosion and consumption of fertilizers and pesticides are considerably less for energy crops than for food crops. On a comparison between coppiced woodland (eg willow trees) and corn, wheat or soya beans, the danger of erosion is 12.5 times smaller, fertilizer consumption 2.1 times smaller, herbicide use 4.4 times smaller, insecticide use 19 times smaller and fungicide use 39 times smaller.[22] The Swedes have found that, if harnessed in the right way, biomass does not result in significant damage to the environment.[23] And that is even without looking at proposals involving short-rotation cropping or at the possibility of doing completely without artificial fertilizers and pesticides. Energy crops thus place a far lesser burden on the land than do food crops.

Harnessing plants as raw materials is by no means just a matter of quantity: it is a matter of quality as well. As previously discussed, unlike fossil resources, the quality of solar resources in their natural state is so high that increasing use of ever more products must necessarily favour or even trigger a shift from mono- to polycultures.

Monocultures occur in food production, which has concentrated on ever fewer species and varieties, principally maize, wheat, rice and potatoes, but they are by no means inevitable in the cultivation of solar resources. In order to exploit the variety of substances directly produced by plants, a great variety of specialized crops must be cultivated. Admittedly, the very low yields for many coveted substances, such as essential oils, are a

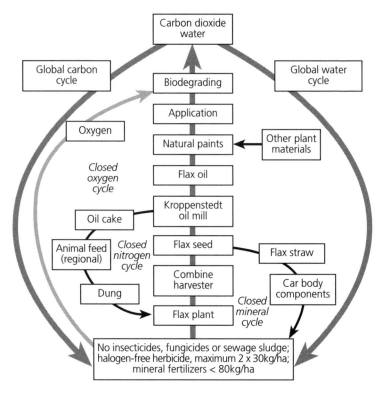

Source: Kroppenstedt oil mill, Germany, unpublished report

Figure 7.3 *The range of applications of a solar raw material*

problem, in that vast harvests are needed to obtain commercially viable quantities. For this reason, solar resources can and should not be cultivated for one specialized purpose alone. The case for organic farming practices improves once the opportunities for and economic advantages of comprehensive multipurpose applications for plant resources and residues are recognized, as depicted in Figure 7.3. Whereas the process of transforming fossil raw materials into chemical products produces toxic waste, the use of solar raw materials opens up the possibility of turning waste disposal costs into additional profit centres. All plant residues not required for the production of a particular product can always be fermented to produce biogas. Productivity considerations alone would lead a chemical industry based on solar raw materials to draw its energy from biological sources. Combining energy generation with material uses simply makes more efficient use of biomass inputs.

The choice is between an industrial focus on a small number of basic products and an agricultural focus on a small number of arable crop species on the one hand, and a multiplicity of basic products and thereby a diverse base of smaller agricultural businesses on the other. The choice is between coarse and fine, single-operator mass cultivation and pluralistic agriculture, monocultural versus polycultural resource production and use. Solar raw materials are vastly superior to a fossil resource base that does not measure up to our current level of knowledge and understanding and which thus keeps production far below the levels that can be achieved.

Chemical products from fossil hydrocarbons are the primary cause of our current waste problem. Breaking synthetic compounds back down into their component molecules is either impossible, or the procedure is complicated and costly. This drastically reduces the scope for recycling, as these substances either do not degrade naturally, or do so only slowly, and so must either be buried or burnt, with woeful environmental consequences. Chemical products produced from plants, however, are not only recycled by nature itself, but their combustion does not release harmful pollutants. This greatly reduces the scale of the waste problem. In addition, people will find waste easier and cheaper to manage. In place of the waste separation regimes in force in Germany and other countries, rubbish will be reduced to two simple categories: easily recyclable metal waste and organic refuse. Recycling of waste itself thus becomes an integral component of a renewable energy system. This stands in stark contrast above all to today's petrochemical products, which usually contain heavy metal additives. In return for marginally lower purchase prices, the consumer is burdened with considerably higher waste disposal costs – yet another example of how the fossil resource industry is hampering the development of productive business models.

In view of the scope for resource substitution, the environmental advantages and the greater productivity gains that solar raw materials offer small and medium-sized enterprises, a blanket rejection of the idea of a solar resource base on the basis of the negative experiences of 'modern' agricultural production would be counterproductive. The existence of dangers (see Chapter 2 for further discussion) cannot be denied. Industrial companies have frequently been known to sacrifice even their own long-term interests for short-term gain, and businesses certainly do not have any consideration for other companies in the industry that are consciously pursuing the non-conventional alternative – quite the opposite. However, anyone who rejects reorientation towards solar raw materials because environmental problems may result must still compare such problems with the consequences of fossil resource consumption. But above all, anyone who rejects a comprehensive transition is leaving the potential of solar resources solely to those who seek to integrate them into the existing fossil-ized structures. That way lies the monoculture – unnecessary, and incapable of tapping the full wealth of solar resources.

Most energy experts, and equally most experts from the environmental lobby who concern themselves with environmentally damaging substances, do not see energy and raw materials as two sides of the same problem. By contrast, the energy and chemical industries, because of their mutual need for fossil hydrocarbons, are well aware of their common interests, even if they do not say so in public. A breakthrough in solar energy use which brought about a fall in crude oil and natural gas consumption would quickly strip fossil petrochemical precursor substances of their current cost advantage over solar materials. The cost of fossil energy would rise if the market for the products of the chemical industry were to shrink. In sum: the more solar raw materials come to replace fossil ones, the more the replacement of fossil by renewable energy sources will be accelerated. This is precisely why the transition towards solar energy and solar raw materials should be seen as a strategic whole. A unitary strategy would make it possible to see through the opposition arguments; solar opportunities would become more clearly visible and tangible, and environmental policy could go beyond the boundaries that have obviously been constraining it hitherto.

As recognition of the wealth of solar resources builds, the economic logic of a solar resource industry will gain the ascendancy. There are two essential maxims:

1   In the manufacture of chemical products, solar raw materials must be preferred to fossil raw materials wherever an equivalent product can be produced from solar inputs.
2   Besides their role as food crops, the use of plants as raw materials has priority over their use as an energy source.

This latter principle does not recant on the goal of meeting energy needs from plant resources, nor does it reduce the potential for energy crops – there are sufficient plant resources to meet the need for nutrition, raw materials and energy. There is also no need to renounce energy crops in order to conserve plant resources for the future: solar raw materials are regenerable. As long as provision is made for future crops, it is possible to switch quickly from one use to the next: from food crops to energy crops to industrial materials, and vice versa. What is necessary – and basic economic management – is to maintain the capacity for agricultural production, from land fertility to biodiversity. Integrated schemes make the most sense, whereby agricultural crops for the different purposes of producing food, raw materials and energy complement each other. The upshot would be a far quicker transition to complete replacement of fossil resources, up to and including fertilizers and pesticides.

The real biotechnology: materials science, not genetic engineering

At first sight, it might seem that industry has already recognized the potential of solar materials. Biotechnology has become a byword for technological innovation and industrial modernization. But it is precipitate to equate – as many do – biotechnology with genetic engineering. Proponents of genetic engineering deliberately encourage this misconception. In the face of ethical reservations and mounting public mistrust, the genetics industry has come to prefer the term 'bio- and genetic technology', or even simply 'biotechnology' – effectively 'greenwashing' genetic engineering. Yet biotechnology is much more than this. It also encompasses fermentation techniques, for example, which have nothing to do with genetic manipulation. The real task of biotechnology ought to be research into the variety of applications to which biological materials can be put.

Reducing the scope of biotechnology to the genetic engineering of people, animals and plants distracts from its real potential: harnessing the wealth of photosynthetic processes that evolution has produced, and which new mutations continue to produce, if we give them half a chance. According to a study by Daniel Querol, every plant has 10,000 genes on average – around 4 billion genes in total for the 400,000 agricultural plants. Every gene has a roughly even probability of mutation, and 400 million years of reproduction behind it. If every year one mutation occurs in every species, then new varieties of each plant are continually being produced, each with 10,000 genes.[24] Natural selection determines how many survive, and what has survived has always been useful. There can be no doubt that the plant kingdom is a cornucopia of ever-increasing riches, constantly in need of new and further research, and as a field of scientific research it offers by far the greatest reward in terms of tapping the true wealth of natural resources.

Concentrating on this field and uncovering the commercial applications is what will really make biotechnology count. The scientific search for what nature offers of its own accord in terms of resource wealth – as measured by the resource needs of industry and agriculture – will be much more profitable than the current focus of biotechnology research: defining a material need and manipulating genes and breeding transgenic plants for as long as it takes to fill the specified need.

It is, however, no coincidence that genetic manipulation is currently in the spotlight. Wresting natural phenomena from their context is hallowed scientific tradition. The chemical sciences, writes Hermann Fischer in his critique of 'hard chemistry', is steeped in the:

> *prejudice that naturally occurring substances neither have the right characteristics nor occur in sufficient quantities to meet*

the major material needs of industrialized societies. It follows from this axiomatic assumption of a qualitatively and quantitatively inadequate natural world that the goal of chemists should be to manufacture their products as far as possible independently of natural processes.[25]

What is unavoidable with petrochemistry is applied to biological materials as method. Gene-orientated biotechnology also suits the marketing needs of the chemical industry, as it seems to offer the fastest means of developing a commercial solution to a specific material need – or at least, far faster than the traditional methods of plant breeding. The current priorities are so-called 'red biotech' – for medicinal, therapeutic and diagnostic purposes – and 'green biotech' for agriculture and food products.

One example is the attempt to make plants pest-resistant by, in effect, writing the pesticide into their genetic code. The rationale for this manipulation is to reduce the need for conventional pesticides while helping in the global fight against starvation. If, on the other hand, biotechnology were to follow the interests of agricultural enterprises rather than the chemical industry, other solutions would also be available, as the Germano-Brazilian agricultural expert and one-time Brazilian environment minister José Lutzenberger never tires of pointing out. For example, in place of pesticides, one might use diluted liquid manure or sugar-enriched ethanol, both of which are inevitable by-products of agricultural production. This method would also strengthen the plants' immune systems.[26] The chemical industry's other objective in developing pest-resistant plants is to make early provision for replacing their petrochemical pesticide products when the fossil hydrocarbons dry up. The global top ten pesticide producers – Ciba-Geigy, ICC, Rhone-Poulenc, the US corporations Du Pont, Dow Elanco and Monsanto, Bayer, Hoechst and BASF and the Anglo-Dutch Shell (in total three German and three US companies) – have a combined turnover on their pesticide products of $5 billion. Some of these companies also figure in the top ten seed merchants, who seek – and have the ability – to leverage GM seed to expand and ultimately monopolize their markets.[27]

The justification given for genetic manipulation is specious, and does not stand up to close scrutiny. The anti-starvation argument has been used and abused time and again in decades past. The stated aim of the so-called 'green revolution' several decades ago was to use industrial agribusiness methods to increase yields across the board. In fact, the effect was just the opposite, although the statistics still trumpet success. Lutzenberger gives one example which is representative of the real impact of agricultural modernization:

> It is argued that the native Indian farmers in Chiapas, Mexico, for whom opposition to NAFTA (the North American common market) is a survival issue, are backward. They produce only two tonnes of maize per hectare versus six tonnes per hectare on modern Mexican plantations. But this is only one side of the coin: the modern plantation produces six tonnes a hectare, end of story. But the Indian cultivates a mixed crop on the same land area. Beans twining round the maize stalks, fruit, pumpkins, sweet potatoes, tomatoes and several kinds of vegetable, fruits and medicinal herbs. He feeds his calves and hens on the same land. He produces a good 15 tonnes of foodstuffs per hectare, and all without commercial fertilizers or pesticides and without the aid of a bank, government or transnational company.[28]

Yet what appears in the statistics is 6 tonnes of maize versus 2 tonnes. Furthermore, the food crops that a farmer relinquishes in order to increase his production to 6 tonnes have to be paid for out of what he does grow. He may have a higher income, but he also pays more for his production and his own food, to the detriment of the environment. Small wonder that the ever lower prices the monopoly purchasers pay him for his crop force him to give up his business and plunge him into destitution – and that the large-scale agribusiness firms either directly dependent on or owned by the food-processing industry flourish.

This agricultural madness culminates in the ever-growing dependency of agricultural enterprises on seed monopolists and patent-owners. The money that used to be spent on pesticide is now spent on pesticide-enhanced seed. Fields used to grow pest-resistant plants cannot then be used to cultivate other plants.[29] Gene patents will ensure that food-processing and chemicals companies are working towards comprehensive patent coverage by offering ever more hybrid plants. These plants have only a limited capacity for reproduction; a farmer wanting to collect seed from such plants must settle for inferior quality. The corporations are also seeking to develop 'terminator genes' which will render plants incapable of reproduction.[30] Farmers will become utterly dependent on seed merchants. This development is driven by purely commercial monopoly interests; it implies an organized campaign to supplant natural varieties, and the end of the free farming community across the globe on the basis of state-patented theft of organisms or parts of organisms, all in the name of the global 'free' market and the global fight against starvation.

On top of that, genetic engineering of plants – from the laboratory to field trials – is associated with high unit costs. Industry is thus strongly motivated to see that the new seed finds widespread use, and all available political strings are pulled to this end. Political institutions, through their research funding, have increasingly endorsed this ruinous form of biotechnology, as Ulrich Dolata documents in his study of corporate strategy, research programmes and technology competition in the area of genetic technology.[31]

The net effect of all this is systematic elimination of biological opportunities, which flies in the face of the wealth of the natural world. A recent report on the biotechnology industry in Germany contains the following passage under the heading '1998 optimism': 'Genetically modified microorganisms, animals and plants will bring about sustained change to agricultural, medical and industrial processes.'[32] This 'sustained change' has nothing to do with sustainable production methods: the current understanding of biotechnology aims to create plants which do not last, and to use fewer, rather than more, varieties. The direction of change must be reversed, and the prime task of biotechnology must be focused research into existing species and their sustainable use.

We need to put nature to economic use to supply our material needs. The problems arise when nature is selectively manipulated without consideration for ecological systems and with no attempt to see, appreciate and understand them in their entirety. We cannot simply apply the methods and procedures of fossil resource use to the economic exploitation of nature. In all probability, genetic research, with its overblown fixation on genetic manipulation with unforeseeable consequences, would proceed with a good deal more caution and with more attention to exploring the countless ways in which existing plant varieties can be used, if the patenting of genes and thus their exclusive use by individual companies were politically off-limits. The patent offices, governments and parliaments who have caved in to barefaced pressure from industrial corporations like Monsanto to allow genes to be patented must bear the responsibility for the competition among companies to manipulate ever more genes and for the biotech industry's headlong rush in the wrong direction.

8

The profitability of renewable energy and resources

AN IMMEDIATE AND comprehensive transition to solar energy must take priority over all other economic considerations. Any further delay will cost society more than it would to make the transition. The quicker and more comprehensively fossil energy and resources can be supplanted by their solar counterparts, the greater the cost saving to society and the less the strain on government budgets threatened by ever higher clean-up costs in the wake of fossil-fuel-induced catastrophe, be it storm or flood damage or regional wars over energy, the growing cost of waste disposal or the cost of maintaining an ever more bloated environmental protection bureaucracy. Almost all environmental damage can be traced to the use of fossil and nuclear energy and fossil raw materials. The greater the investment in solar resources today, the lower the costs imposed on tomorrow.

The longer this transition is postponed, the more costly it will be to implement, as the external costs of the fossil resource base are swelling exponentially.

Yet despite modern society's propensity for squander and waste, the energy debate is being conducted in the most pernickety and parsimonious of terms. There must be a thorough cost–benefit analysis, we are told, to determine whether we can 'afford' a sustainable energy supply. At least, prominent politicians and businessmen seem to assume that society is inclined and compelled to think in such penny-pinch-ing terms. Energy for tomorrow must not cost more than energy that cannot last beyond today! The use of such inappropriate yardsticks is symptomatic of the moral prostration of politicians, and the expression of the shameless self-aggrandizement of the business world over the present and future victims of a wantonly destructive energy system.

The debate about the cost of renewable energy illustrates just how far we have to go before we achieve the levels of civilized behaviour in this area that elsewhere we regard as a matter of course. Nobody has the right to simply leave their rubbish lying in the street – patently, the cost of proper disposal must be paid, which in Germany adds up to over €200 ($182) for a four-person household. Energy waste in the form of emissions, by comparison, can simply be tossed out to pollute the atmosphere and environment. Even if renewable energy did cost more than is actually the case, the correct response would be to pay up without demur and alter our spending priorities accordingly. This principle must take precedence over all cost considerations in respect of renewable energy. Those who can see the dangers of continued reliance on fossil resources, but who nevertheless maintain that averting these dangers must be a commercially viable proposition; those who maintain that the introduction of renewable resources must not infringe competition rules to the detriment of fossil resource suppliers: such people will run the economy into the ground within the first half of the 21st century. Ultimately, they are a danger to themselves. The broad-brush approach to the issue of renewable energy and the blanket dismissal on the basis of allegedly too high costs reflect a desperate search for excuses to justify the blatant failings of the fossil energy system.

Irrespective of the basic principle that we must afford investment in the solar future and take it forward as a matter of priority – before road-building and military projects, before subsidies for old structures and, as individuals, before buying expensive cars or foreign holidays – optimal costing of solar investments is indispensable if best use is to be made of the financial resources available. For various reasons – not just to do with avoiding environmental damage – the cost considerations for solar energy and materials differ from those that pertain to fossil resources. The many factors just discussed which contribute to the potentially far higher productivity of solar resources will only be reflected in cost calculations if profitability is not calculated simply by applying the metrics developed for conventional fossil fuel power generation to solar resources. As long as the wrong measures continue to be used, solar resources will trade below their real commercial value.

Whose costs? Why solar and fossil resources cannot be compared on the basis of economic efficiency calculations

Chapter 5 described the misleading way in which the economic viability of an energy source is assessed, whereby energy costs are equated with energy prices with no consideration for any productivity gains. Equating price with cost implicitly assumes that individuals and companies have no control over their energy use, which is patently absurd. Electricity suppliers have scope for making more efficient use of energy in the power stations they operate, both by improving plant efficiency to produce more electricity per unit of primary energy and, if possible, by finding a market for the spare heat energy. The consumer has scope for making more efficient use of energy through more efficient boilers and cars, better insulation, energy-efficient appliances and by cutting down on energy-consuming activities. This is also the tack taken by the numerous proposals for reducing the use of conventional energy.

Productivity gains are also at the heart of the so-called energy contracting model. Under this model, a client company outsources the management of its energy systems to a contractor, for a period of perhaps ten years. The contract price is set at the level of existing annual energy bills, multiplied by the length of the contract. The contractor then implements cost-cutting measures at their own risk, including the necessary capital investment. The greater the cost reductions that can be achieved without compromising service quality, the greater the return to the contractor. It is simply good business practice to treat energy as a variable cost. This elementary principle has as yet had little influence on the general debate on energy costs, as it does not fit the energy industry business model of constant supply. Least-cost planning analyses,[1] which compare the cost of investment in energy conservation with the cost of building new power stations, also show that the former, as a rule, offer the greatest return to energy suppliers.

The cost flexibility of renewable energy, however, goes far beyond the cost reductions that can be achieved using fossil energy. Whereas nuclear and fossil energy have just one cost/benefit ratio, renewable energy sources open new economic possibilities: solar panels doubling up as cladding for buildings, agricultural residues as an energy source, and many others. These provide additional opportunities for cost-cutting, and investing in renewable energy can even bring in additional revenue for the operator.

All that is needed for this to happen is to make the cognitive leap to see that the operator of a PV installation can step out of their consumer role to become an independent energy entrepreneur. As such, they can calculate their costs exactly as would a business when buying new equipment or machinery, and deploy their investment as flexibly and therefore as profitably as possible. A computer, for example, can be a tool for writing or revising documents, performing calculations, drawing, copying, translating, transmitting and receiving information and information storage. If a profit-seeking company were to use this computer solely for writing documents which had no need of subsequent revision, the costs in comparison to the classical typewriter would be so high as to make the computer unprofitable. Profitability increases rapidly, however, paying back the purchase cost many times over, as soon several or all of its functions are put to use. The economic value of an investment grows with its flexibility. Costs are not immutable, but vary according to the specific circumstances of the investment.

The energy cost comparisons that we usually get to see are not as sophisticated as this. As a rule, they simply compare the unit capital cost of the investment, on the basis of average cost per kWh of generation capacity. In the case of PV and wind, the comparison is subject to the proviso that, due to the discontinuous availability of sunlight and wind, installed plant cannot operate continuously and thus the effective average annual output is lower than that of a conventional power station. In consequence, the real cost of energy output from PV and wind power plant is higher than a simple comparison of capital cost would suggest. Calculations like this are fair enough; it is just that allowances for discontinuous operation in the comparisons between conventional electricity generation plant found in studies and publications are conspicuous by their absence – as if all conventional power stations operated continuously, which, of course, is not the case. The actual annual output from renewable energy plant can be calculated on the basis of the figures for average insolation and wind strength at the particular location, whereas the annual operating times for individual conventional power installations are not known and are concealed to boot. Where figures can be concealed, published costings can be massaged to suit. This alone calls into question profitability figures calculated solely on the basis of a comparison of capital costs. The cost of a product depends on how and to what end it is used.

There are, however, many other factors that call the standard comparative evaluation of profitability into question. The first step in assessing the profitability of a solar power installation is always to determine the purpose of the investment: classical energy supplier or private operator? An energy supplier sells energy to customers at a price that takes account of all its costs and profit expectations. A supplier can also be a producer, but the cost of producing energy is always lower than the cost of supplying it, because this includes the cost of transport and distribution, marketing and invoicing. Defining the term more precisely, the supplier is the agent who sells to the end user. In the electricity industry, this role is usually filled by the distributing companies; in the case of fuel, it is the garages; for heating oil it is the oil merchant. In the case of renewable energy, however, the separation of roles between energy supplier on the one hand and energy consumer on the other either does not apply, or can become increasingly blurred, given the possibilities described in Chapter 6.

In the area of electricity generation, operators of renewable energy generating plant can:

- be an integral part of the overall electricity supply system, whereby they simply act as a producer selling electricity to a retail supplier;
- take on the role of retail supplier in addition to functioning as producers, and sell their electricity to end-users;
- be self-sufficient, whereby they use the plant to supply their own needs and become energy-autonomous;
- be self-sufficient and produce electricity for a retail supplier at the same time, by selling energy surplus to their own needs; and/or
- supply both themselves and other end-users simultaneously.

Each of these cases displays a different cost/benefit relationship. Profitability calculations will also be different, depending on the purpose to which the installation is or can be put, with variations in the flexibility available for cost-cutting. There are specific calculations to be performed for every supplier or consumer set-up, which in part go beyond those needed for fossil energy supply. Multifunctional applications of solar technology also allow further cost reductions, which do not find their way into calculations based purely on energy costs.

Operators of solar installations have, nevertheless, hitherto mostly confined themselves to calculating the benefit in terms of filling their own needs. Before the energy markets were opened up to competition, independent operators were prevented from supplying other users. Also, affordable storage technologies, which would bring cost-effective self-sufficiency in electricity within reach, have yet to appear on the market. Consequently, there are currently two aspects to electricity generation from solar sources: supplying the needs of the operator on the one hand, and feeding surplus energy into the grid on the other. Already, this involves two kWh calculations for the same plant. The price paid by the grid (in Germany regulated by a system of guaranteed minimum payments) gives the operator a lower return than does using the electricity to reduce the amount they take from the grid. The more an operator can supply their own needs, the lower their costs.

As this example indicates, it is cost avoidance that is the most important factor in calculating the profitability of a solar installation. Renewable energy offers a raft of opportunities to avoid costs, opportunities not available to users of conventional energy.

Cost avoidance: economical application of solar resources in a nutshell

How renewable energy can be used to avoid costs is the crucial question for widespread use of renewable energy. It is a question which everybody must ask themselves: developers of manufacturers of solar technology, in order to enhance the marketability of their products; operators of solar plant, in order to increase their scope for investment and maximize their economic benefit; manufacturers of electrical appliances, architects and property developers, in order to improve their products and find new markets.

There is one cost which is avoided with all solar energy installations, with the exception of biomass combustion: there are no running costs. This is a well-known fact, yet profitability calculations do not always give it due consideration, in particular through long-term calculation of the savings on energy bills. These savings should be incorporated when assessing loans for solar installations, just as in the case of mortgages for buying or building property. Unlike a normal business loan, the payback periods for mortgages are longer, and the savings on or income from rent are included in the financing calculations. Building societies and property financing companies were founded for this purpose: banks have specialist departments, and these also include any government funding in their calculations. Such factors have so far rarely been considered when evaluating investments in renewable energy.

Long-term cashflow analysis over the design lifetime of a product is the only adequate method for calculating cost, and not just for renewable energy. The trend in business towards decreasing use of such analysis, because short-term payback periods are increasingly becoming the standard yardstick for evaluating investments, even to the point of governing liberal economic policy, deprives the economy of its prospects for the future. As short-term calculations come to dominate, the long-term cost to the economy mounts. In no field have the negative consequences been experienced more immediately than in construction. Houses must be durable goods because of their high costs alone. The economics of modern construction gives primacy to the capital costs of construction, which are also the crucial factor in the awarding of contracts through competitive tender. The consequence has been a rapid rise in the incidence of structural problems appearing soon after construction, due to the use of substandard construction materials, with the result that demolition becomes necessary only two or three decades later. The public sector sets a bad example, due to budgetary regulations which – in Germany at least – impose short-term expenditure constraints, rather than accounting the operating costs for a building over a period of at least 20 years. The attitude is very much one of 'après moi, le déluge'. Long-term operating cost calculations for buildings make solar construction techniques all but essential.

Solar building cost calculations

The extent to which costs can be avoided by employing solar energy in building construction is largely determined by the following:

- The extent to which the demand for electricity and heat energy can be reduced by siting the building to make maximum use of the sun's heat and light, through insulation and heat reclamation or by using devices powered by solar panels.
- Whether solar panels and solar collectors are 'bolted on' or integrated into the fabric of the building, thereby replacing other components and reducing costs. PV panels or solar collectors which are simultaneously part or all of a roof or cladding save on the cost of conventional roofing or cladding. The comparison per kWh of electricity or unit of heat input with a conventional electricity supply or heating system is no longer the deciding factor, but rather the cost comparison between a building component including its energy output and an unproductive component.

- Whether the solar installations are only add-on components of the energy system, with the implication that all
  the costs of conventional energy systems are also still
  incurred, or whether they in fact constitute the entire
  energy supply, thus rendering conventional energy equipment, including boilers and grid connections, unnecessary.
- Whether all the solar energy collected can be used with the
  aid of built-in storage capacity, or whether the surplus is
  lost or must be passed on.

The more functions that can be performed by solar energy and
solar surface technology within the building, without recourse
to conventional energy plant within the building or to external
energy supplies, and the more other costs can be avoided
through use of solar technology, the quicker the use of solar
energy in buildings will become so profitable that the costs are
not just on a par with those of a conventional building, but
can even outbid them. Solar costs become solar profits. The
spectrum of possibilities is broad, and there are already numerous practical examples.

The restored Reichstag building in Berlin, for example, is
equipped with its own cogeneration plant running on vegetable
oil. This unit is capable of supplying all the building's heating
and electricity needs. The only reason that this does not happen
in practice is that in summer, the heating system does not run
at capacity, and consequently the supply of electricity is insufficient. By contrast, when the system does run at capacity in
winter, the output of electricity is sometimes enough to allow
energy to be fed into the grid. With the addition of a cost-
effective storage system, the building could become completely
self-sufficient in electricity.

It is already feasible with existing technology to heat a
moderate-sized house from solar irradiation alone, if the building is sufficiently well insulated and constructed to maximize
solar gain, such that solar collectors and heat stores are only
needed to supply a small proportion of the total heating need.
Using solar heating to supply all a building's heating needs,
rather than just as an adjunct to a conventional heating system,
is the more economic proposition. The conventional heating

system can be dispensed with, and there are no more heating bills. In the case of the on-site biomass cogeneration plant discussed previously, it would become possible, with the aid of an electricity storage system, to dispense with the grid connection entirely. This would save the cost of the connection and the electricity supply, while also running a car at a fuel cost below anything else on the market. The economic viability of an energy storage system depends on the cost of the storage system versus the savings on the additional external energy supply that would otherwise be necessary.

A solar village of several hundred houses designed by the architect Rolf Disch is currently under construction near Freiburg. The houses in this development produce more energy from the sun than they actually need. The cost was calculated on the assumption that the development would otherwise need to be served by conventional energy. In the conventional case, the total cost of a single house came to €343,687 ($305,500), whereas the solar version costs €307 ($270) less! Reduced energy loss and additional energy gains bring the annual bill to €17,851 ($15,870), versus €18,433 ($16,380) using conventional energy. After 15 years, once the investment in the solar plant has been amortized, the savings over conventional energy will be €2045 ($1800) annually – and that is without taking account of likely rises in conventional energy prices, perhaps to twice their current level. Moreover, this solar village does not exploit all the technological possibilities that will be available in future.

The cost of solar energy

The inadequacy of costing energy supplies on the basis of the capital cost per kW of capacity, as described above, does not just stem from the assumption that the plant – with the exception of PV and wind power – will always run at capacity. All the sundry operating costs are also assumed to be identical, which is patently not the case. The cost of electricity generation normally derives from the following factors:

Capital costs
Initial planning
Purchase of land
Cost of power station
Installation costs
Ancillary buildings
Grid connections
Technical monitoring and quality control
Cost of capital

Operating costs
Fuel
Personnel
Insurance
Maintenance

If we compare this with decentralized renewable energy generation plant, then a number of these costs no longer apply:

- Minimal initial planning requirements, and none at all in the case of small plant.
- No land purchase is necessary for photovoltaic installations and internal cogeneration facilities in buildings. For wind power, land is only a cost factor where the plant operator is not the owner of the land (usually an agricultural enterprise). The small footprint of a wind turbine has a minimal impact on the agricultural value of a plot of land, and farming can happily continue around the base of the tower.
- Domestic installations do not need dedicated grid connections where connections already exist and are paid for by the operators of the installations in their domestic electricity bills. The same applies to new developments, where the cost of a grid connection must be paid in any case – that is, unless the intention is to create energy-autonomous dwellings.
- Micropower plant incurs a lower cost of capital than large power stations because of the difference in lead time. Large power stations take a long time, usually several years, to

construct. No power can be generated, and thus no repayments made, before construction is complete. Installation of micropower plant, on the other hand, is a matter of mere hours, which means that power generation and principal repayments can start at once.

It is not just the fuel savings on wind, PV and hydropower installations that vastly lower the operating costs of renewable energy generation plant: personnel costs are also either non-existent or very low, especially in the case of PV. Monitoring is continuous and usually performed by the individual operator.

The advantages of these installations make them ideal for domestic use. But they also offer an unmatched benefit to power companies as well. Although there would be a need for staff to monitor the plant, distributed power generation from renewable sources is a concert of innumerable independent modules, which means that the risk of costly misjudgements of capacity requirements is virtually eliminated. No longer needing to build large power stations, companies can expand capacity module by module to meet demand, with zero lead time. Where there is excess capacity due to changes in demand or because demand has been overestimated, there is no need to shut down an entire power station: it is sufficient merely to take a couple of modules off-line. Stranded investments can therefore largely be avoided, or kept to a minimal level.

It will already be clear that all calculations that go beyond the capital cost of the technology favour renewable energy. This is most true of PV, which incur virtually no secondary costs. Investment analyses which do not take account of this are incomplete and superficial, stamped from the accounting mould of the existing energy industry. Such analyses also lead to spurious conclusions. The only financial disadvantage that still obtains in respect of renewable energy is the still relatively high cost of the generation plant. This cost, however, can be brought down ever further through technical improvements and industrial mass production, as the history of every technology of the past two centuries has shown.

It is, however, also clear that the cost of renewable energy falls when generated at the domestic level, rather than by power companies. As long as the large power companies refuse to switch to renewable sources, domestic installations do not represent a second-best solution, but rather the superior business model. This becomes all the more clear with the shift from domestic power generation to complete energy autonomy, ie, dispensing with the grid and the associated costs. Standard domestic customers in the EU currently pay €0.10–0.15 (5–7¢) per kWh. By contrast, the published cost of generating electricity in fossil fuel and nuclear power stations is only €0.02–0.04 (1–2¢). If the published cost is accurate, then generation costs account for only 30 per cent of the retail price. The remaining 70 per cent goes on the grid infrastructure and the provision of adequate capacity.

There is thus enormous scope for self-sufficiency and independence from the grid. Deciding whether to move towards self-sufficiency and grid independence is a question of comparing the cost of the generation plant and power storage system together with operating costs with the price of grid electricity. If the capital cost – written down over several years – plus operating cost falls below the price at the meter, then solar energy self-sufficiency is the financially superior investment.

The cost of agricultural energy

The economics of solar energy could be the saving of many agricultural enterprises – an insight that is key to the development of agriculture as a whole.

Agricultural enterprises consume large quantities of energy, including indirect energy in the form of hydrocarbon fertilizers and pesticides produced by the chemicals industry from fossil fuels. This is the greatest single drain on the financial resources of farming businesses, and consequently one of the two main causes of business failure in the agricultural sector. The other main cause is the downwards pressure on farm-gate prices exerted by wholesale traders in agricultural produce and the food-processing companies through the 'free' market.

Spending on fertilizers, pesticides and energy makes up 30 to 35 per cent of total agricultural expenditure, which shows up in the statistics as 'industrial pre-processing'. In the case of arable farming, however, the proportion is far higher than this, sometimes rising above the 60 per cent mark. According to a 1987 study, expenditure on pesticides alone is DM175 (€90; $80) per hectare for wheat, and as much as DM200–300 (€100–150; $90–$135) per hectare for potatoes or sugar beet.[2] Aggregate expenditure on direct and indirect energy by the German agricultural sector was DM11.3 billion (€5.8 billion; $5.14 billion) in the financial year 1985–1986.[3] The figures are clear: if agricultural enterprises could be spared these costs, their long-term financial viability could be assured, alongside more opportunities for independent marketing direct to the end-consumer. Whereas the latter ultimately depends on the political framework for the market in agricultural produce, farming businesses do have ample scope for breaking away from the use of fossil energy and thereby increasing their returns.

For decades, the maxim was growth or bust – raise yields to maintain revenue. And the only way to do this, the farmers were told, was massive use of chemical fertilizers and pesticides. The outcome was devastating: yields were increased but incomes fell, and ever more family farms were compelled by apparently insufficient size to sell up. Between 1950 and 1970 alone, as this process was getting underway, production – now reduced to single crops – grew by 70 per cent, but consumption of direct and indirect energy and thus energy expenditure also grew several times over. The increased production costs meant that the increased yields actually lowered farming revenues, in a kind of one-off anti-productivity boom. All the indications are that it was not necessarily the fertilizers and pesticides that made these increased yields possible, but the use of agricultural machinery.

In a book entitled *Regenerating Agriculture*, Jules N Pretty compares the profitability of farms using agrochemical production methods with those forgoing the use of artificial chemicals. Moreover, he compares farms from the same regions, ie, with very similar climate and land quality. The result is striking. Yields are in many cases about the same, but incomes are higher where agrochemical methods have been abandoned, which also benefits the environment.[4] Obviously these farms have not forgone fertilizers and pesticides entirely. Instead, using the methods detailed in Chapter 7, they have manufactured their own from agricultural residues, thereby revitalizing the natural circular flow of nutrients. What makes this all the more astounding is that in only a few cases did this approach also include the substitution of fossil fuels with self-produced biological replacements, for which biogas or vegetable oil would be the obvious candidates. This provides scope for significant reductions in another cost factor, roughly equivalent to the savings on fertilizers and pesticides. To realize these reductions, the motors and machinery used in agricultural production would need to be converted to run on vegetable oil or biogas.

Replacing direct and indirect fossil energy with in-house products does make agricultural production more labour-intensive, but this is outweighed by the increased returns that result from the cost savings. As production costs fall with the shift away from 'industrial pre-processing', this also has far-reaching consequences for the scope for cultivating biomass as an industrial raw material and energy source in the future. The agricultural sector gains new opportunities for direct marketing of these resources at lower prices than before, thereby accelerating the move away from fossil hydrocarbons in the chemicals industry and the move away from fossil energy in society as a whole.

PART IV

TOWARDS A SOLAR ECONOMY

Dissenting opinions notwithstanding, a return from fossil fuels to solar energy is inevitable. The only open question is whether this happens in time to avert the impending environmental, economic and political disasters of fossil-fuel consumption, and for new economics to bring new ecological stability. The first half of the 21st century will decide the fate of human civilization. Assuming each generation of responsible economic agents spans around 30 years, it will be the next two generations who will have to make this process of eco-industrialization happen. The dangers are too great to risk further procrastination. It is high time the world drew back from the precipice to which fossil fuels have brought it.

Ever more frequent and ever more devastating environmental catastrophes are a warning we dare not ignore. Such disasters are happening earlier and more often than even critical climate researchers dared to predict: 707 in 1998 alone. The *World Disaster Report* published by the Federation of International Aid Agencies lists around 60,000 lives lost; the El Niño weather disturbance alone cost 21,000 lives; Hurricane Mitch in Central America 10,000. The floods in China affected 180 million people; there were also a further 240 storms and 170 floods affecting a total of 200 million people. The droughts resulting from climate change in Indonesia led to fires in rice paddies and forests which caused a pall of smoke that darkened the skies over southeast Asia for weeks. Who could fail, in such circumstances, to have an inkling of how the end of a civilization with a pyromaniac energy system might look?

Environmental disasters forced 25 million people to flee their homes; Munich Re estimates the bill to be $90 billion, versus $30 billion in 1997.[1] The consequences for the refugees, from cultural uprooting to social destitution, are incalculable. That most of these catastrophes result from the meltdown of the fossil energy age is something that only the terminally blinkered can now contest. The countdown has begun, and the time bombs are not just ticking more loudly now: many have already gone off.

Dissenting opinions are now rarely to be heard. Nevertheless, despite all the warm words for renewable energy, the business and politics of energy supply carry on regardless. Actions in practice work more against the alternative that in its favour. For top of the agenda in politics and business is not the restriction, but rather the encouragement of conventional energy consumption. The most obvious option, using tax increases to restrict consumption across the board, has been inadequately enacted. Instead, national and international political initiatives are deliberately seeking, in the expectation of general approbation, to lower the cost of energy. Dismantling global trade barriers and new market structures in the electricity and gas industries have resulted in a fall in prices that will lead to an unfettered continuation of the orgy of energy consumption for a long time to come. No attempt has been made to counter these falls with appropriate rises in energy taxation. It is impossible to conceive of a wider gulf between understanding and action, at a time when a sustainable and risk-free solution to humanity's perennial problem of energy supply has come within reach.

Across the world, the last market barriers facing suppliers of conventional energy are falling. Just as with the telecommunications industry before it, the area monopolies of the gas and electricity industry are being rescinded. The electricity oligopolists and monopolists, though, are no less powerful for it. New entrants do now have access to once-closed market segments, which at first sight is a welcome development, as the ossified structures of past decades can now be broken open. Proponents of green alternatives have – all too precipitously – welcomed the opening of the energy markets for precisely this reason. However, the market power and financial strength that the energy companies were able to build up under the system of regional monopolies gives them a head start in the liberalized energy markets. While the ex-monopolists certainly compete with one another, they also have the capacity to effectively stymie new entrants offering greener products. The most likely result will be a 'clearing' of the market – a market for ever fewer, ever larger electricity companies. De jure, there may be more opportunity for independent supplies of green electricity from local micropower plants; de facto, the development has made life difficult for municipal corporations and for those trying to bring renewable energy onto the market.

The behaviour of governments in pursuing energy market liberalization with the sole object of securing lower energy prices, while simultaneously encouraging the process of concentration in the energy industry, is inexcusable. Scarcely a merger or acquisition has gone by in the electricity industry that was not welcomed or even initiated by the governments concerned. Energy markets that do not give precedence to renewable energy, price cuts without effective environmental taxation, the encouragement of mergers and acquisitions at the cost of local solutions: such policies represent an intellectual step backwards and are all but irresponsible.

To bow to the premise that energy can only be supplied within structures predicated on the consumption of fossil fuels is to don an intellectual and practical straitjacket – in science as much as economics, in politics as much as culture. It becomes impossible to conceive of a world in which solar replaces fossil energy, or that this could bring advantages for civilization as a whole. It also becomes hard to design an appropriate strategy, for which the aim must be to harness the power of solar energy through an optimal structure for this energy source. Only in this way can solar initiatives be liberated from the snares and obstructions of the fossil-fuel economy.

Nevertheless, the view that solar energy and resources should be exploited within this structure and integrated into the existing system is widely held, even among supporters of the solar alternative. Adapting to structures more suited to the natural, technological and economic potential of solar energy, on the other hand, means thinking outside the box. This is easy, and yet hard at the same time: easy, because second thoughts cost nothing and anyone can have them; hard, because it means, in effect, having to think in a totally new direction. The 'grand strategy' for the transition to a solar global economy is a transnational social project that has to be realized by innumerable agents and in countless steps, small and large. Every step has a value, in that it brings the shift to a new economic base closer. Some courses of action will meet obvious limits; other cross boundaries and break new ground. These are the courses of action we must pursue. In *A Solar Manifesto*, I set out a broad palette of possible initiatives, which I will not repeat here.[2] Instead, I address two questions:

1   Which initiatives are capable of side-stepping or overcoming existing structural obstacles in order to accelerate the take-up of renewable resources? The pace has patently been too slow thus far; if it is to be stepped up, all policies have to be evaluated for their potential scope.
2   Which policies are required in order to dethrone the fossil resource industry from its current hegemony? The desire to effect change cannot be reconciled with a hands-off policy on precisely those structures of the fossil resource industry which are the mainspring of its enduring expansion in the energy, raw materials and foodstuff sectors.

It is a question of which are the fastest roads, and these will not necessarily be the most direct or the most comfortable. The way must be marked so that ever more agents can follow it. It is not just a matter of what is simplest to achieve and therefore the pragmatic response now. The modern fixation with what looks possible today dulls the eye for what will become indispensable – and then achievable – tomorrow or the day after.

9

Exploiting solar energy

ANY STRATEGY FOR an across-the-board introduction of solar resources will be inappropriate or limited in scope if it does not also provide for the strategies of individual energy companies. There is a wide variety across the industry in terms of the way that these companies are responding to change and to new challenges. Some are looking to the oil fields of the Caucasus rather than those of the Middle East; others are seeking to tap 'non-conventional' fossil fuel reserves, whatever it may cost; others still are refocusing from oil to gas; some are diversifying their businesses into areas other than energy. A few are thinking further ahead and beginning to diversify into renewable energy by starting up appropriate subsidiaries. Notwithstanding these differences, there are four broad patterns which can be discerned:

1   Heightened industrial concentration, with political backing, through strategic mergers and aggressive competition against smaller firms. The public is left with the impression that energy firms are becoming increasingly dominant, and that the need to accommodate to their wishes is greater than ever. In reality, these firms are mustering their strength for a last life-or-death struggle.
2   Attempts to meet political demands for global climate protection in a way that does not call the future of the nuclear and fossil energy business into question. This is the motivation for proposals such as tradable global emissions rights: 'Yes there must be change, but not in my back yard!'
3   As energy companies have realized that their strategy of categorical rejection of renewable energy (apart, of course,

from the highly profitable hydropower dams that have long since formed part of the energy mix) is no longer tenable, they are attempting to ensure that renewable energy reaches the market on their terms.

4   A drive to bring energy suppliers from different sectors together to develop integrated services, albeit in a way that suits the energy companies, ie, within existing hierarchical structures.

It is important to bear these trends in mind when evaluating renewable energy strategies. As desirable as it may be to compel energy companies to do their bit for the transition to a solar global economy, through public pressure, persuasion or governmental fiat, it is unrealistic to expect companies to willingly substitute renewable for fossil energy — ie, to act against their own interests. Even if the factors discussed in Chapter 2 make it impossible to control renewable energy supplies in the same way as fossil or nuclear energy, who sets the pace and whose interests prevail remain questions of fundamental importance. It has been and continues to be independent agents without wider connections who lead the charge: grass-roots organizations, individual operators, new companies, municipal utilities, politicians. It is these agents who have mounted the public education campaigns and prepared the market for solar technology. Rather than relying on corporations and governments to take the reins, these organizations need support in their work. 'Wir haben verstanden' (we have understood) — that was the slogan adopted by the German Social Democratic Party (SDP) after its poor performance in the 1999 elections to the European Parliament. The current situation in the energy markets shows just how few really have understood — particularly in respect of the electricity market, where the lure of cheap electricity from conventional sources is being held out to tempt consumers to buy into the destruction of their own future.

The transition to a solar global economy cannot be achieved without the combined activities of local and independent agents and innumerable small investors. They are what is needed if the technologies and proposals for exploiting solar resources are to be developed to the point where they become obviously more affordable to the public than conventional energy supplies. Only then will the shift to solar energy really get under way. Without political action to revoke the privileged status of conventional energy suppliers and to overcome the market advantage enjoyed by the established energy businesses, however, there is still a great danger that it will take too long to set the stage for this historic energy sea-change. The criteria for evaluating individual initiatives are clear: the ultimate aim of all policies to promote solar energy and bring it to the market must be to turn the economic advantage of solar energy – very short or non-existent supply chains – into strategies that can further accelerate the pace of change.

The role of capital allowances – and their problems

Renewable energy entered the market via financial support from governments for private investment in solar energy technologies. Public subsidies are still the primary form of political support for initiatives on the ground. There were and still are good reasons for this, as energy consumers, faced with the difference in cost between conventional energy supplies and the capital cost of renewable energy equipment, needed financial incentives to encourage them to invest in solar power. The palette of financial aid instruments ranges from direct subsidies covering a set percentage of the cost of equipment, through to tax breaks and low-cost loans at subsidized interest rates.

As important as such programmes and incentives are in kick-starting new trends, it is important not to gloss over their faults. In many cases, they are or have been little more than gestures, and some have done more harm than good. That goes for short-term programmes with small budgets, whose funds are often very quickly exhausted. Applicants who miss the boat are put on the waiting list for the next financial year, with the consequence that individual plans get put on the back burner. Announcements of funds that then fail to appear also have dire consequences. For example, in 1996, the Italian government announced a programme to mount PV on 10,000 roofs, but after three years there had still been no movement on it, and in the meantime the still weak PV sector – without which a programme like this cannot even be implemented – had collapsed. With friends like that, renewable energy has no need of enemies.

Subsidies are also a two-edged sword. One the one hand, they represent an incentive; on the other, they have led to investment in solar energy becoming a by-word for financial pump-priming, with the result that almost every investment decision is taken subject to the availability of state support, even in the case of applications that are financially viable on their own merits. This dependency culture has become a psychological barrier to an across-the-board introduction of solar energy. In the absence of subsidies, there is a widespread knee-jerk assumption that solar technology is too expensive. Such reservations are even held in respect of built-in solar panels in electrical appliances, despite the fact that no additional cost would be involved, and that no purchaser would even ask about the cost of what would be but one component among many. There are any number of buildings with features, such as specialist cladding, which increase the cost of construction, but if such features also function as solar energy systems, they suddenly become too expensive and require subsidies. According to a study by Eurosolar on water taxis in Venice, solar-powered boats would cost no more than diesel-powered ones, but with the added benefit of reducing water and air pollution in the city while protecting canal-side buildings from vibrations.[1] Solar-powered taxis have also failed to appear due to a lack of subsidy.

I do not document such behaviour in order to disparage pump-priming programmes. That I was instrumental in enacting what have so far been the biggest two such programmes in the area of renewable energy is testimonial enough to my innocence in this respect. One of these was the DM200 million (€102 million; $91 million) pump-priming programme for renewable energy launched by the SDP/Green Party coalition government in 1999; the other was the so-called '100,000 solar roofs programme', which I initiated and which was also launched in 1999 with a budget of DM1 billion (€500 million; $455 million). This latter aims to see 100,000 rooftop PV systems or 300 MW of PV capacity installed over six years;

the incentives are interest-free loans and a grant covering 12.5 per cent of the cost of the plant. The intention is to stimulate mass demand and, with the transition to mass production, to lower unit costs.[2] The programme also meets the essential requirements for financial incentives to:

- provide long-term funding to prevent hiatuses in the expansion of the market;
- have a budget large enough to forestall the development of waiting lists that put a brake on individual initiative;
- aim at a reduction in the unit cost of the product, by triggering competition among manufacturers to achieve favourable prices and productivity gains;
- try to develop a self-sustaining market, which means that there must either be a follow-up programme or market regulation to promote further expansion.

The energy levy act passed by the Swiss Nationalrat (the National Council, the lower house of the Swiss Parliament) in 1997 comes closest to satisfying ambitious aspirations for solar energy technology. The legislation envisaged a levy on all conventional energy, and the money so raised – around SwFr1 billion (Swiss francs) (€684 million; $581 million) – was to be ringfenced for programmes to support renewable energy and energy conservation initiatives. The levy was to have remained in force until 50 per cent of Swiss energy demands were met from renewable sources.[3] Unfortunately, there then followed a two-year tug-of-war between the two chambers, with the Ständerat (the Council of States, the second chamber) demanding a lower rate of levy. The legislation was finally defeated in a referendum held in September 2000, amidst popular unrest over the then comparatively high oil prices.

Note: the scale of the problem requires far more public money to be made available for pump-priming, but only on the basis of the criteria set out above. As it is neither possible nor conceivable to reshape the entire energy system through subsidies alone, public money can only be a temporary measure to kick-start a new developmental trend, and not a substitute for more far-reaching measures on market regulation and other policies.

Tax-exempt status for solar resources: overcoming the legitimacy crisis of environmental taxation

As convincing as the case for eco-taxation – higher taxation for environmentally damaging forms of energy combined with reductions in other taxes – may be, it stands in stark contrast to the sluggish progress on implementation. Indeed, the idea has actually become less popular, for which the most important reason is that the increases and reductions in taxation do not affect the same people or even occur at the same time. The people who bear the burden of additional energy taxation are not always the same as those who benefit from reductions in other taxes. In consequence, acutely felt rises in energy taxes are rarely seen in proportion to reductions elsewhere; rather, energy taxation is seen in the context of exhortations by parliaments and governments for self-restraint among the general population. (In the UK, for example, the Climate Change Levy introduced in April 2001 recycles the revenue from additional energy taxation in the form of a reduction in payroll taxes. However, not all heavy users of energy are also major employers – and vice versa.) The practical upshot of the aim of stimulating investment in energy efficiency is that the money must be found for additional investment on top of higher energy prices, if total energy costs are to kept in check. This opens the eco-tax to accusations of placing additional burdens on individuals at a time when individual incomes are falling in any case, as a consequence of more general economic trends. This became particularly clear in Germany during the 1998 election, when the Green Party repeated their demand that petrol duty should be gradually raised to give a final price of DM5 (€2.6; $2.30) a litre. This demand, part of Green Party policy for many years, met with widespread resistance as never before. When the Greens argued that the arrival of super-economical cars long before this point would mean that even with a price of DM5 a litre, fuel costs per 100 km would be no higher than previously, the response was that not everybody could afford a new car.

There has also been increased resistance from the business community. As global competition has grown tougher, declarations that eco-taxation damages international competitiveness are finding more sympathetic ears. The motivation behind proposals to introduce such taxation at EU, rather than national level, is often to see that it never hits the statute book. Although it would be desirable to see such measures implemented at the international level as soon as possible, it is no excuse for postponing action at the national level. After all, to combat proposals for EU-wide energy taxation, industry can always fall back on the argument that if such taxes are to be imposed at all, they need to be imposed on a global scale, or at the very least, in all industrialized countries. It almost goes without saying that proposals for global taxation stand no chance of getting off the drawing board.

In the light of these experiences, it is past time that the cycle of delayed and watered-down legislation were broken. That means framing eco-taxation proposals in such a way that they do not become just another tax rise, raising fears of threatened livelihoods and excessive red tape. Instead, they must be founded on a clearly articulated strategy to drive nuclear and fossil fuel energy out of the market, to be replaced by renewable energy. The objective of eco-taxation must not be to rein in energy consumption as a whole, but to stimulate the switch to renewable sources. This does not mean that energy prices must sky-rocket: the duty on conventional energy should be set at a level which makes renewable energy the cheaper option. Eco-taxation will garner even more support if at the same time renewable energy is made tax-exempt – ie, if alongside the increased burden, there is a tax-free alternative which people are encouraged to use.

If fuels from renewable sources can already be brought to the market for under €1 (88¢) a litre, assuming tax-free status, then the tax of fossil fuels need only be sufficiently high to make them appreciably more expensive – ie, a little more than €1. That would be sufficient to trigger the rapid displacement of fossil fuels from the market. The giant leap for civilization away from oil – the largest single sector within the energy industry – can be one comparatively small and comprehensible step for the consumer.

In order to free environmental taxation from the stigma of being socially regressive, the glaring injustices in the energy taxation regime must be rectified by rescinding the tax-exempt status of certain classes of activity. In particular, this means an end to the tax-free status of shipping and aviation fuel (for which there are also important economic arguments, which will be discussed in more detail in Chapter 10) and, within the EU, the exemption of the oil-processing (ie, chemicals) industry from oil duties, the devastating consequences of which were dealt with in Chapter 7. These tax breaks keep consumption of fossil energy high and prices artificially low, even outside the areas directly subsidized. Reduced total sales of fossil energy will automatically result in higher prices, thus accelerating the process of their replacement. Increasing taxes on conventional energy while preserving the tax-exempt status of high-demand industrial sectors would be a glaring contradiction. Tax schemes that do not resolve this contradiction by cancelling such subsidies lack credibility and focus.

'Green' taxation schemes must also be made more equitable by ending the blanket exemption for business use. Exempting business exempts the sphere of activity which has the greatest scope for effecting change. The necessary investment can be offset against tax. At the very least, exemptions must be made conditional on proof that the company in question has exhausted all its options for rationalizing its energy use (for example, by contracting out its energy management, as described in Chapter 8). This proof could take the form of an energy audit. The ten-year period mentioned in Chapter 8 was not an arbitrary choice: most asset-management contracts are of a similar duration.

There are other forms of tax-break which meet the criteria I describe, such as a 50 per cent or full rebate on VAT for installation of solar plant or for the electricity it produces. Such a measure would probably have an at least neutral net impact on state finances: while it would lose out on one tax stream, the boost to solar technology would result in increased revenue from other tax streams – for example, additional income taxes resulting from new jobs. Zero-rating for VAT could also be used to provide an incentive to switch from fossil to solar raw materials, and so accelerate their market penetration.

The current 'three-litre car' debate in Germany illustrates just how much green tax proposals have missed the target of moving away from fossil fuels. In this instance, the German government proposed to use eco-taxation to incentivize purchases of new hyper-economical cars. In view of the world-wide growth in car sales, however, fuel-efficient engines do not represent a real green alternative. The sheer number of new cars would negate any gains from more efficient engines. Half the fuel consumption times double the number of cars equals no change in the high global consumption of fossil motor fuel. Zero-emissions vehicles are a much more convincing target – ie, cars running on renewable energy sources. Zero-emission vehicles would be just as quick to develop and bring to the market, and would even make car travel cheaper again if fuels from renewable sources were tax-exempt. The pace of market penetration would be far quicker than for fuel-efficient vehicles (whereas the cost of fossil motor fuel would remain high). Fuel duty could be reinstated on bio-fuels at any point, once fossil fuels had vanished from the roads. Once the supply chains have been dismantled, there would no longer be any going back to fossil energy. All the indications are that the more radical strategy of making the transition to renewable energy would not only be a better solution to the problem, but would also be a more popular one, and thus more politically feasible.

Possibilities and problems in the market for green electricity

The principle that underlies legislation opening up electricity markets – which in the EU are governed by the 1997 Common Market directive on electricity – is the functional separation of generation, high-voltage transmission and retail distribution through the low-voltage local grid. Initially, this was a shock to the system for the electricity industry;[4] it has since been welcomed. It was also welcomed, however, by most proponents of renewable energy because of their negative experiences with regional monopolies. Yet the two sides have diametrically opposed strategies and expectations: the existing companies, acting on the basis of the market power acquired during the monopoly years, are looking to expand their markets; suppliers of electricity from renewable sources, by contrast, are looking for unimpeded access to the market.

In some countries, though, the first blow to the regional monopolies had already been dealt, in the form of feed-in legislation securing access to the grid for operators of renewable energy plant by setting minimum prices that the grid operator had to pay. In Germany, I helped initiate and drive forwards both these necessary measures. The question is, what will be the long-term impact on the substitution of fossil by renewable energy, against the backdrop of current developments in the electricity market?

The enduring conflict surrounding grid feed-in laws

The first grid feed-in laws for renewable energy, passed in Germany, Denmark, Spain, Italy and Greece,[5] guaranteed independent operators access to the grid and a price for their electricity sufficient to provide a secure investment environment for private operators. This alone was not enough to ensure that renewable energy penetrated the market, however: private investors in Italy in Greece were required to meet technical specifications which effectively blocked the practical application of the law. It is no coincidence that these are countries with single state electricity companies. In Germany, Denmark and Spain, by contrast, the new laws sparked a slew of private investment by independent operators in a relatively short space of time. Above all, it sparked investment in windfarms, for which the guaranteed prices were sufficient to ensure profitable operation. Grid feed-in legislation has thus proved to be the most successful policy instrument for bringing renewable energy onto the electricity market. Two of the preconditions for this success were the existence of public awareness of and backing for renewable energy in the countries concerned, and an electricity supply not based on a single state utility, so that the idea of independent electricity suppliers can quickly take hold. That said, it is remarkable that there is so little enthusiasm for independent electricity supplies in a country like Italy, which is known for its cultural emphasis on individuality.

Right from the start, grid feed-in laws met with opposition from the electricity industry, and this opposition grew as more private operators entered the market. The established electricity companies have been using scare tactics to fight their corner, claiming that industry and consumers would be burdened with unacceptable costs. They argue that though they burn less fuel when electricity from renewable sources is fed into the grid, they still have to maintain all their capacity to cover for other times. This argument assumes that production, transmission and distribution forms an integrated business, although such integration was officially legislated out of existence by the new laws governing the electricity market. In other words, their position is unlawful. Within the framework of the new legislation, additional costs to the consumer may only result from the difference in price between the average cost of supplying electricity at the level of the (low-voltage) distribution grid and the legal price minima for renewable sources. The government could in any case legislate to redistribute any additional costs among all the grid operating companies, in order not to disadvantage those who happen to have to purchase particularly large quantities of electricity from renewable sources, for which a higher price must be paid than for electricity from other sources. Grid feed-in laws incorporating minimum prices thus represent the easiest means of continually expanding the proportion of electricity generated from renewable sources – always assuming that governments and parliaments have sufficient political will to enact such laws or maintain them on the statute book, and to ensure compliance. If precedence is accorded to generating energy from renewable sources, then the contribution they make can grow year on year, with a concomitant fall in the demand for electricity from nuclear or fossil sources. If electricity from renewable sources is also tax-exempt or becomes so, and duty is only payable on conventional energy, then the additional cost to the consumer will automatically be lower. There will in any event be a need for further study to determine the true level of any additional costs. What is certain is that they will be considerably lower than has been argued, and in some cases possibly non-existent.

PV electricity is always generated during the times of the day at which demand reaches its peak and electricity therefore is at its most valuable. Electricity suppliers typically purchase peak-load electricity for prices between €0.13 and €0.26 ($0.11–$0.23). There also tends to be more wind during the day than during the night, which means that windfarms also generally produce more valuable electricity fetching a higher price. This suggests than price minima can be differentiated to reflect the actual average cost of electricity during the different demand periods, which would especially benefit PV. German renewable energy legislation which came into force in April 2000 specifies a minimum price of €0.51 ($0.45) per kWh, and a yearly depreciation rate for new installations of 5 per cent. Assuming that the resultant widespread take-up of PV and further technical development produce lower prices year on year, then the price of PV electricity would eventually fall to the level of the average cost of peak-rate electricity. From this point on, prices set according to actual market values would be sufficient to cover investors' costs.

Ironically, the very success of grid feed-in laws renders it unlikely that they will be introduced universally or be allowed to stand for long. The greater their success in increasing the proportion of electricity generated from renewable sources, the greater also will be the opposition to them, and the electricity industry is more than capable of being obstructive. Despite the legal separation of generation, transmission and supply in the electricity industry, electricity companies are actually tightening their grip on all three functions, whether directly or indirectly, by acquiring more and more local distribution grids. The aim is to reach the end-consumers, regardless of the fact that the new rules were supposed to disentangle production and distribution and thereby put an end to the situation in which producers could abuse energy grids to secure a monopoly position.

The established electricity companies have a common interest in preventing the uncontrolled entry of renewable energy into the grid. Renewable energy is an irritant because the growing contingent of renewable energy suppliers are not under the direct control of the established electricity companies. The lobbying efforts of the latter have proven so successful that the success of grid feed-in legislation in the above-named countries has attracted little attention elsewhere. In its White Paper on renewable energy in 1997, the EU Commission did put forward a directive on grid feed-in, with a minimum price set according to the average cost of electricity supply up to the 'city gate' of the local distribution grid, plus an environmental premium of 20 per cent.[6] The European Parliament also passed a motion supporting the directive.[7] The Commission Directorates-General responsible for drawing up directives for the energy market and for competition policy, however, have so far remained stubbornly opposed to implementing these proposals – just as they have ignored the clauses in the Common Market directive on electricity that explicitly allow for preferential treatment at the national level of electricity generated in an environment-friendly manner.

Attempts by companies to have grid feed-in laws replaced by a quota system form another component of this defensive strategy. Electricity companies would be obliged to purchase a proportion of the electricity they sell from renewable sources, this proportion being set by the government, which would raise it from time to time. The quota would be filled by a process of competitive tender, the idea being to bring market forces, and the resultant pressure to raise productivity, to bear on renewable energy. Such proposals, however, overlook regional differences in the availability of renewable energy, whose profitability depends not just on the productivity of the plant, but also on the local wind conditions and insolation. The consequence of systems of quotas and competitive tender would be a scramble for the best geographical locations and for the renewable energy technologies that currently offer the most favourable production costs – at the moment, primarily wind. The supply of good locations would quickly be exhausted, especially with the local opposition that results from disproportionate regional concentration of such plant. Other electricity suppliers would be forced to use more expensive suppliers with less favourable locations as a consequence of losing the competition for the best locations. The market distortion argument would make an immediate reappearance, especially from the mouths of those who currently propose quotas as an alternative to legislation on grid feed-in.

At the very least, every subsequent increase in the quota would be the subject of intense controversy, as the problem of different prices in different locations would become more acute with each rise in the proportion of electricity coming from renewable sources. Pressure for enlarged quotas could only come from independent operators not encumbered by existing conventional capacity. Yet the lack of such encumbrances is a precondition for the rapid expansion of the market for renewable energy. If a quota system were required to ensure that investment in renewable energy did not concentrate on particular technologies such as wind power, which currently has a cost advantage over other renewable sources, then the quota would have to be divided into subquotas, which would turn the system into a bureaucratic nightmare. And as the quotas would also be filled using remote production capacity, renewable energy would become increasingly dependent on the national grid. Quotas can also be side-stepped by awarding the contract to a supplier without requiring delivery. This has occurred time and again in competitive tender processes under the British non-fossil-fuel obligation, which is a model for quota systems elsewhere. Quota systems therefore cannot be an alternative to grid feed-in rules with minimum prices. Electricity companies extol quota systems because they offer a way to resurrect their regional monopolies: the first quota would be filled in-house, and further increases resisted.[8]

There is no obvious sensible alternative to minimum price legislation. The Renewable Energy Act passed by the German parliament in February 2000, which replaces the existing grid feed-in legislation, is also based on minimum prices. The difference with the new act is that the payment mechanism takes full account of the unbundling of the local grid. Bridging finance for electricity from renewable sources comes from the national grid, following the statutory price minima. The national grid in turn apportions the cost to all electricity suppliers according to the proportion of total electricity supplies coming from renewable sources. All suppliers, and by extension all consumers, thus contribute to the financing of investment in renewable energy in proportion to the share of renewable energy in national electricity supplies. The result is a sort of sliding quota with no limit on the quantity of renewable energy that can fed into the grid at statutory minimum prices. This adheres to the market principle as far as it can be applied to renewable energy. But it is still environmental considerations that dictate the primacy of renewable energy, thus bringing about a fundamental shift in policy in an era of electricity market liberalization.

Independent markets for green electricity and the intermediary problem

An independent electricity market consists of direct contracts between green electricity suppliers – either an appropriate broker or a direct producer – and customers. The contract covers the supply of electricity generated solely from renewable sources, with electricity from fossil fuel cogeneration plants representing a halfway-house.[9]

Suppliers of green electricity have begun to appear as electricity markets have been opened up to competition, especially in countries which have seen wide-ranging public debate on environmental alternatives, and where these have broad public support. Such suppliers seek to attract customers who are prepared, whether for reasons of principle, caution or public image, to pay higher prices for green energy. Surveys have shown that large swathes of the population are willing to pay higher prices,[10] and there is also a precedent in the form of the market for organic produce. The market for green electricity may even grow faster than the organic produce market, as the consumer is not faced with a new set of purchasing decisions every shopping trip, but simply has to sign on the dotted line when they change their supplier. The enthusiasm of some new suppliers of green electricity is so euphoric as to make all other options seem beside the point. This optimism rests on two expectations:

1 continuing development will bring the price of green electricity down year on year, thus lowering the price differ-

ential with conventional electricity and making it easier to grow the market for the alternative – especially if renewable energy is granted tax-exempt status; and

2   grid operating companies will operate a non-discriminatory pricing structure, thus ensuring a level playing field for all electricity suppliers.

But these expectations are far from secure, as the bombshells dropped by the electricity companies have shown. It is highly doubtful that the price differential between green and brown electricity will move in favour of green electricity in the near future. A price-cutting war has broken out in the European electricity market. The large corporations are holding four cards that give them considerable scope for action. Firstly, they have a large pool of power stations for which the depreciation charge has already been paid, and which can therefore be operated very cheaply with minimal additional investment. Secondly, electricity can be bought cheaply on the international market, especially from Eastern Europe. Thirdly, the established companies were able to build up a large financial cushion during the regional monopoly years. Fourthly, they have considerable scope for cutting staff. This is why electricity prices have recently been sinking by double-digit rates – it is indirect evidence of earlier excessive profits at the consumer's expense. These trends will lead to a widening of the price differential for the foreseeable future, thus reducing the ability of green electricity to penetrate the market.[11]

There are also question marks in respect of transmission charges and discrimination – whether a non-discriminatory regime can be achieved as hoped, and whether such a regime can remain free of barriers thrown up by established electricity companies. There can be no question that the functional separation of production, transmission and distribution is not being observed in practice. Even where power companies have been divided into generation and grid-operating companies, attempts to use the grid to secure the position of large power stations continue, and transmission charges are the weapon of choice. ETSO, the Association of European Transmission System Operators, the peak association for European operators of high- and medium-voltage grids, was founded in July 1999 in order to exert pressure on the Commission and the governments of the Member States to accept the existing transmission agreements within the industry, and not to issue additional directives or create an independent watchdog or competition authority with the power to intervene against discriminatory behaviour. Even if the rules on transmission are not obviously discriminatory, the major grid operators have ample scope for placing technical difficulties in the way of unwelcome competitors.

Transmission charges are the principal bone of contention. Some green electricity companies reject transmission charges outright, on the basis that these have already been paid by the customer before the switch to green electricity, through the charge for the provision of supply capacity contained in the electricity price and through the connection charge. According to this line of argument, an additional transmission charge would represent a triple payment by the green electricity consumer.[12] It would therefore make more sense to contract only for the additional cost of investing in renewable energy. The investment would be used to install new plant, and the electricity produced would be fed into the grid as usual and paid for in line with the statutory minima. As electricity is an intangible product, this is no different in physical terms from a contractual relationship between the green electricity supplier and the customer. Electricity always goes into a general pool; it is impossible to determine the origin of what comes out the other end. For such business models to work, however, there must be a system of statutory minimum payments in place.

In the face of competition from suppliers of green electricity, electricity companies have started to create their own green electricity subsidiaries. Minimal additional investment is required, because established companies have a legacy stock of hydropower capacity, which they can take out of their existing supply mix in order to sell it as green electricity. This enables them to undercut any other green supplier. Despite their enormous competitive advantages, however, they undercut other suppliers only by a very small amount, thus making additional profits without altering the overall balance of renewable energy in the system. This tactic enables electricity companies to exploit the positive attitude of green electricity customers towards renewable energy to rake in additional profits, improve their public image and block or neutralize the threat of independent green suppliers.

The only counter to such tactics is a certification system for suppliers of green electricity. Certification must be conditional on the credibility of the supplier and on proof that the electricity comes from new plant. The German green electricity certification association Grüner Strom-Label, which is supported by, among others, the environmental organizations BUND, Naturschutzbund, Deutscher Naturschutzring, Eurosolar and IPPNW, the energy consumers' association and Verbraucherinitiative (a national consumer action group), therefore refuses certification to green electricity companies with considerable holdings in electricity companies that continue to include nuclear and fossil energy in their strategy and whose behaviour hampers the development of renewable energy, despite statements to the contrary.[13]

The outlook for the market for green electricity thus depends on whether the behaviour of the suppliers is transparent and whether the transmission conditions directly or indirectly favour established electricity companies, be that by multiple charges for the same service or by high charges for high-voltage transmission. The grid has proved to be the power companies' most lucrative asset, and it has turned them into an economic force to be reckoned with, despite the existence of state regulators. The potential of the green electricity market lies not in a few large suppliers, but in many regional ones. Local markets also do more justice to the unique nature of renewable energy as a local energy source. In this case, there would be no need for transmission charges, because these really would have been covered by the connection charges paid by both the supplier and the customer. In the case of supply contracts between suppliers and customers within the local distribution grid, (German) municipal utilities rightly do not impose further transmission charges. The correct approach for renewable energy and for a distributed energy supply would be different access charges for external suppliers and suppliers operating within the local grid, according to the principle of 'from the local area to the local area', or 'from the region to the region'.

Green suppliers and municipal self-sufficiency

The lack of transparency in the way electricity prices are determined within the grid is what enables companies to confuse the debate on grid feed-in laws and the real costs of supplying electricity from renewable sources. The mechanism by which prices are set and supply contracts agreed is an impenetrable black box, which is why the grid-based electricity industry can quite happily make unverifiable claims. The pitfalls of the green electricity market also have to do with the problems of the grid, despite the fact that a decentralized generation industry does not need a national grid. It has already been established that the best way to exploit the potential of renewable energy is not by replicating the centralized grid. Instead, renewable energy needs its own supply structure based on high-performance electricity storage.

The argument against integrating renewable energy into the existing grid does not stem from dogmatic rejection or demonization of the industry, but from the simple fact that this is not the most economically efficient way to harness technologies that can be deployed anywhere and whose multi-functionality offers multiple economic benefits. The decisive element in any strategy for renewable energy will be an understanding of its very different supply-chain requirements. It is important to be aware of the structural conflict between the necessarily hierarchical organization of nuclear and fossil fuel energy supply and the different requirements for renewable energy, and not to muddy the waters. Whereas the established energy companies are becoming diversified but nevertheless highly centralized energy suppliers, the goal for renewable energy should be a trans-sectoral distributed energy supply system. The foundations of any energy system are its sources and consumers of energy. Extended supply chains were necessary and a centrally managed system possible with nuclear and fossil fuel energy because source and consumer could not be brought together within the same locality or region (except, that is, in those regions where primary energy is extracted). With renewable energy it becomes possible to co-locate source and consumer, thus rendering transmission over a high-voltage grid unnecessary. It would be like taking an unnecessary detour via a private toll-road. A distributed system, on the other hand, would revolutionize energy supply and open the floodgates for renewable energy. It is also the only way out for municipal and regional electricity suppliers, including the municipal electricity utilities. This is the path they must tread, lest they find themselves fighting a last-ditch battle against industrial consolidation – a fight that, without an integrated energy base, they can only lose. The range of agents who would participate in a localized energy system include:

- classical energy consumers, who buy all the energy they need (including motor fuel) from the local or regional supplier;
- individual energy users who generate part of their own electricity and buy the rest on the market;
- the completely self-sufficient, who store their own energy for later use;
- the self-sufficient who become suppliers themselves by selling their excess power to the local or regional supplier; and
- the diversified green energy supplier who buys others' surplus energy for redistribution or storage, and who operates the local electricity and gas network.

If demand for a centrally managed energy system falls and eventually disappears as the renewable energy sector expands, then the individual consumer need no longer pay for it. Specialized conventional energy suppliers will give way to a decentralized energy system. It is the responsibility of the operator of the local distribution grid to optimize the productivity of privately operated plant by redirecting surplus energy to other uses. Both the local grid operator and the operators of private installations, including the energy self-sufficient, increase their productivity by exploiting the multifunctionality of renewable energy to the full, and by adding new modules as the technology matures. Their role is thus not just purely economic.

Distribution has always been the preserve of municipal enterprises, which must now become integrated energy businesses. Municipalities must ensure that the local grid remains under their control, and that they regain control if it has already been sold. The local grid, however, is more than just a delivery service for electricity: its function is to store all surplus power from local generation plants, both those under the control of the local supplier and privately operated individual systems. This integrated municipal corporation would also provide all the other network carriage services – ie, sewerage, gas, distributed heat and water networks. It would also use these networks to offer internet connectivity. All these services can increasingly be operated in parallel, thus saving infrastructure and maintenance costs. The services described properly belong in the public sector because there are no competing networks for electricity, water, gas or distributed heat, which means that market distortion is not an issue. The only questions are:

- Who controls the networks? Private monopolies, or democratic institutions that can deliver cost transparency and prevent discriminatory behaviour? We would do well to remember that at the dawn of the modern urban age, it was private businesses who demanded that municipal infrastructure should be run by the local council, with equal charges for all users: for reasons of market economics, no monopoly enterprise should be allowed to compete simultaneously in other sectors.
- Will local distribution networks remain integrated, so that they can be used more productively for multiple supply functions?

Privatizing supply networks will not bring about competition. Competition comes from different firms producing different products in different ways, and from the separation of production from the use of networks. In order to ensure more productive energy use, a municipal energy corporation would provide energy storage systems both for municipal power plants and as a service to independent operators. It would also produce and sell biofuels – vegetable oil, biogas, hydrogen, methanol, ethanol or gasified biomass from regional sources – and provide a network of garages.

A municipal energy corporation would thus provide storage services for all forms of energy, assuming that individual operators do not store their own energy. It would enter into partnership with local agriculture and forestry companies, buying raw biomass and converting it into electricity, heat or heating or motor fuel. It might buy agriculturally produced biogas and sell it as motor fuel or for generating electricity and heat. It would sell the residues from its own biomass power plants and gasification plants or the fermented biomass from its own biogas plants to agricultural enterprises as fertilizer, and, with further processing, as chemicals for pest control. The principal obstacle to municipal cogeneration plants – the difficulty of finding customers for the distributed heat – would no longer apply, as any excess heat could be converted into electricity, and any surplus electricity could be converted into other energy media.

A municipal energy supplier would always have enough stored energy to be able to meet demand spikes from local production. There would no longer be a charge for conveying energy from local suppliers to local consumers, because it would be covered by the connection charges. Individual suppliers of green electricity would therefore be able to find the markets they need within the local area. The local network of resource-processing enterprises would form the new basis for the functional separation of roles within the industry; with each additional module, their independence from the conventional fuel, gas or electricity supply chain grows; the system becomes more effective and more economic. Businesses and property owners have the option of installing their own generating plant, moving further towards self-sufficiency as the necessary technologies become available. Cross-substitution of multifunctional energy storage systems to iron out demand fluctuations would put an end to conflicts over capacity. In addition, the municipal corporation could also offer contracting and consultancy services.

Any responsible local councillor with a clear understanding of these possibilities should feel obliged to set this process in train, which means above all regaining or retaining control of local supply networks and pushing the integration of energy functions to the top of the local political agenda. The result would be more jobs for local people and a stronger local economy.

The path I describe leads from external supply to regional, municipal and individual self-sufficiency. For many individuals, this will mean autonomous energy supplies; for society it means no long-distance power cables, apart from the few lines needed to bring power from large dams and windfarms. As the centralized gas and electricity industry loses its purpose, and municipal power supplies and energy self-sufficiency expand into more and more domains, the high- and medium-voltage cables will gradually disappear. Pylons will no longer march across the landscape, more than compensating for any decline in landscape quality due to wind turbines. The big energy suppliers would dismantle the pylons themselves, as the maintenance costs would no longer be worth their limited use. Private operators of PV, wind, hydro and biogas plants will – supported by grid feed-in laws – either become partners of the municipal energy corporations, which in turn will no longer have difficulty finding customers for their energy, because surpluses can be stored for later use – or they might become direct suppliers of green energy, equipped with their own storage facilities and supplying customers round the clock. Perhaps they will also run garages for vehicles running on electricity, compressed air, fuel cells or hydrogen-driven internal combustion engines.

Just as electricity need no longer be supplied through centrally managed supply chains, so too with motor fuel. This is no utopian vision, but a real possibility: centrally run energy suppliers will always have higher overheads than local suppliers, even with a common distribution infrastructure. Synergies achieved through consolidation cannot achieve the same economic results as a local network, because the latter can deploy storage and conversion technologies far more flexibly and make more imaginative use of its assets. Not least, a local

Hierarchical divisions within the conventional energy system between source and consumer

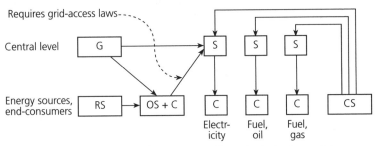

Integrated hierarchical energy supplies – the industry model

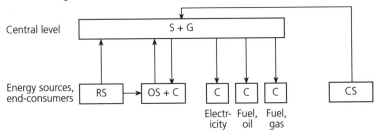

Integrated distributed energy supplies using renewable energy

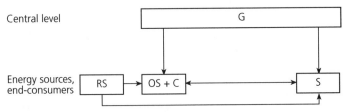

Note: RS = renewable source; OS + C = own supply and consumer; C = consumer; CS = conventional source; S = supplier; G = generation plant

Figure 9.1 *Energy supply structures incorporating renewable energy*

network has no vested interests or large investments to service. Integrated solar energy self-sufficiency is not susceptible to external interference.

The way to achieve this vision is to use local planning powers to bring all the local supply infrastructure together under one network operating company. Existing municipal power companies that generate electricity and distribute electricity and heat should be broken up into network-operating and energy-supply companies. The latter would provide

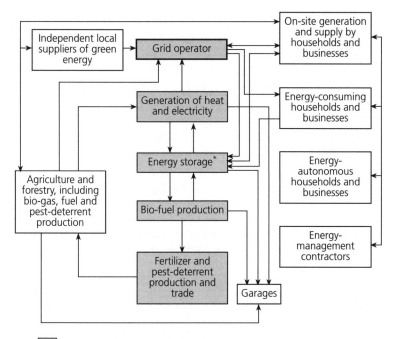

Note: ▢ Possible roles for a municipal energy holding company
* Including locally collected organic waste, bio-gasification, hydrogen production

Figure 9.2 *Model for the future: municipally/regionally integrated energy supply incorporating renewable energy*

electricity, heat and motor fuel, and maintain storage facilities to this end. This structure provides optimum scope for cross-substitution and multiple asset use. Energy production and supply can become cheap and environmentally benign.

The ability of a local network to meet peak power demands in a cost-effective way offers the greatest scope for taking the wind out of the sails of the established electricity industry. Although established companies can also make use of new storage technologies, the much larger scale of their capital installations makes them that much more costly to upgrade. Their inability to put the waste heat from large power stations to economic use, due to the prohibitive cost of constructing the infrastructure for distributing heat, illustrates the extent of the difficulties these companies face. The basic advantages of decentralized local solutions are much lower infrastructural costs and the unique ability to expand production in small and manageable modular steps. Figure 9.1 contrasts three basic models for energy supply: the current hierarchical structure of conventional energy supply; the consolidated system that existing companies hope to achieve; and a local integrated network. Figure 9.2 presents the forward-looking model of an integrated municipal network in its economic context. The result is a public–private partnership at the local or regional level which can avoid conflicts over capacity with a fundamental rethink of the market mechanisms, looking towards to a future total energy service industry (Heinz Ossenbrink).[14]

Creative destruction in the energy industry and the transformation of the resource industry

The transition towards a solar global economy follows a completely different path to that taken by the conventional energy industry. In an age where the conventional energy industry is playing an even larger role, and especially where transnational mergers threaten to make the industry all-powerful, even many supporters of renewable energy think this unrealistic. The sheer concentrated power overawes political institutions, the public and even environmental activists. Many thus look to cooperation with the energy industry, rather than competition. After all, new initiatives suggest that energy companies are finally getting the message, after years of ignoring and obstructing renewable energy; we still need the grid infrastructure, and their financial capital is urgently needed to bring about the desired environmental sea-change – surely cooperation is the only realistic option?

Arguments that appeal to realism always beg the question: which reality? And whose? Physical laws excepted, all 'realism' is a subjective and usually incomplete assessment. It is realistic to note that renewable energy can no longer be ignored, and that energy companies are also beginning to take an interest. It is realistic to see that the fossil energy industry is continuing to expand, and that the trend towards consolidation is accelerating. When dinosaurs mate, their offspring are not pussycats. It is realistic to realize that the scale of the problem makes it irresponsible to continue expanding the renewable energy sector at the current snail's pace.

Joseph Schumpeter, one of the greatest economists of the 20th century, asserted that economists were making a fundamental error by only looking at data from one point in time, 'as if there were no past and no future, believing then that they had understood all there was to understand'. The replacement of fossil fuel by renewable energy is what Schumpeter would have described as a 'process of industrial transition, which inexorably revolutionizes the structure of the economy from within, inexorably destroys the old structure and inexorably creates a new'. This is what he termed 'creative destruction'.[15] However, this process requires larger economic agents than are to be found among the suppliers of renewable energy, now or at any time.

At first, the energy industry confined the renewable energy boat to port; now they are looking to board and take over the cabins. Now that the boat is finally underway, they want to control its speed and have their hands on the tiller. But the way the energy industry is bound up in its own supply chains should lead us to expect it to sound the 'all-stop' at the very latest when its own structure comes under threat. At that point, it would only be necessary to keep the industry on board if its expertise were indispensable and if there were no other suitable major players. But there are.

Allies and co-sponsors for renewable energy are to be found in all those industrial sectors that stand to gain from a solar energy supply, and who can scarcely lose. A broad range of enterprises in industry, in the trades (manually skilled self-employed professions) and in agriculture has an objective interest in helping to shape the transition; it is just that most of them have yet to recognize this subjectively. They underestimate their scope for action and overlook the opportunities to be gained by driving change. In some cases, they would be able to rely on their existing business model with only a few technical adjustments; in some cases, they would even be able to count on their established markets. In other cases, there would be a need to find new markets and customers: companies in the business of supplying large capital installations to large clients would need to move into supplying smaller installations to a vastly enlarged customer base.

The electrical and electronics industries are just as well placed to leverage renewable energy technology as are suppliers of machinery and plant and the construction materials industry. Car manufacturers can expand their markets by diversifying into motors for distributed solar energy systems with applications well outside the road transport industry. The agricultural machinery industry, which has for years been losing many customers to agricultural business failure, can expect demand for biomass harvesting equipment to breathe new life into their old markets. Manufacturers of large-scale power plants will need to look for new customers, but the vast numbers of micropower installations that will replace the large power stations will allow them to grow their businesses. Suppliers of construction materials will retain their existing markets through the transition to solar materials, but will need to change their suppliers.

There will inevitably be losers: in the oil, gas and coal industries, among manufacturers of extraction and energy transmission plant, and in the ranks of cabling firms and operators of power stations and grid infrastructure. These firms have the option of diversifying, shifting from fossil energy supply to solar cell or wind turbine manufacture, although they have no more aptitude for solar technology than any other industry – in most cases, probably less. The technological expertise and existing client base of manufacturers of motors, machinery, plant, electrical appliances and construction materials give them a much better chance of becoming a driving force for renewable energy. They are not hampered or trapped by legacy investments in the fossil energy supply chain, and technologists, electronics experts, farmers, civil engineers, architects and craftspeople, meteorologists, biologists and chemists will be of more use to a firm seeking to exploit renewable energy than the knowledge base of the conventional energy industry in geology, extraction and pipeline technology, conventional power stations, high-tension transmission and high-voltage transformers, etc. Once the interest of the technology industry in accessing new markets is taken into account, along with their greater experience in the marketing of smaller technology goods, they can be seen to have more motivation, to be better equipped and therefore have more scope than classical energy companies for succeeding in the solar global economy.

Industry must throw off the physical and intellectual chains that bind it to the energy business, or else squander its greatest opportunity for future success. Only for the existing energy industry will the transition to a solar energy supply mark the end of the road – something that industrialists, like the general public and many political activists, have yet to realize. In a solar-powered future, every industrial business is also an energy business.

The unique role of the chemicals industry was discussed in Chapter 1. Its close functional links with the oil and gas industry explain why it has so far failed to capitalize on the opportunities renewable energy affords it. Yet the contribution of the chemical industry is of central importance to much solar technology – for example, the manufacture of silicon or other solar cell substrates, battery technology, electrochemical and thermochemical energy storage media, insulating materials. Even more significant is the contribution it could make by replacing fossil and mineral raw materials with solar substitutes in, for example, plastics, dyes, paints, varnishes and medicines.

Unlike the energy industry, the chemical industry is not faced with dissolution, and it has certainly developed an acute awareness of new opportunities, as the strong interest in biotechnology shows. Trying to tailor new ventures to fit existing production processes based on fossil resources, however, is a dead end. If the established chemicals industry does not make the leap from fossil hydrocarbons to solar materials, this task will fall to new chemicals enterprises. Tapping not just solar energy, but ultimately the materials potential as well, will lead to a complete shake-up of the supplier base. More and more suppliers will join the existing few as new applications are discovered for ever greater numbers of plants. The result will be a decentralization of the raw materials industry.

As industrial companies come to recognize and capitalize on their opportunities, new alliances will be formed: between electronics and glass, between the building materials and electrical industries and manufacturers of solar collectors and PV, between motor manufacturers and suppliers of chemical equipment. New groupings will form as old alliances dissolve; as the fossil industrial web unravels, so too will the power structures it sustains. Some existing giants will retain their size; some small PV and wind turbine manufacturers will grow to the proportions of today's car manufacturers. There will be a rash of new specialist biotech enterprises. Since the dawn of the Industrial Revolution, new mass-market technologies have been bringing about economic upheavals – but none compare with the dissolution of the fossil resource base and its global supply chains.

A solar resource base is no impediment to consolidation among manufacturers of solar technologies, but it does rescind the globalization imperative currently driving the resource industry. The transition to a solar resource base will loosen the fossil clamps on the global economy. It also signifies a fundamental turning point in economic history: away from the inexorable trend towards an ever-decreasing pool of mega-corporations, and towards sustainable smaller and medium-sized business forms, which will be and must remain embedded within a regional context, and whose business models may even prove more effective than those of large structures. This is the most comprehensive and most significant structural change in the history of the global economy, and thus also the most controversial.

Hard roads to soft resources

Conflict for conflict's sake is unproductive for all concerned. Consensus for consensus's sake is a pyrrhic victory. Horses for courses: each may be the appropriate response to a given situation. Social conflict is necessary where illegitimate interests stand in the way of a legitimate, uncontroversial and universally desirable aim. Consensus must be sought where common ground can be found between the aspirations and interests of different groups, or if conflict between them would drag in the entirety of society, including disinterested third parties.

Unfortunately, it is entirely legal to exploit resources whose processing imposes hefty burdens on society and disrupts ecosystems. It would even be legitimate, if there were no other way to satisfy the material needs of humanity. But the existence of other possibilities has collapsed the social legitimacy of the nuclear and fossil energy industry. At the very latest, this will be abundantly clear to the next generation. Continuing to cling to the current system without good cause will render conflict over the introduction of solar alternatives inevitable. Any consensus that postpones the shift away from fossil fuels in order to protect established interests does untold damage to society, whereas conflicts that advance the alternative are beneficial – even if not always to the individual protagonists.

There is not even any need to pick a fight with the nuclear and fossil fuel industry: it is already happening, with a stream of new offensives from the major agents in the industry, from international conflicts over access to resources, through to systematic attempts to block successful initiatives to build a market for renewable energy. Calls from the energy industry to seek consensus whenever their interests are threatened are pure hypocrisy.

The question is thus not whether conflict is to be desired, but whether proponents of alternatives to the nuclear and fossil fuel resource industry themselves recognize and have the courage for the existing conflict. Just how doggedly the apologists of the conventional energy industry are prosecuting this conflict can be seen in the innumerable reports and announcements portraying a distorted picture of renewable energy and the state of the technology, not infrequently produced with the aid of generously remunerated scientists. The aim is to create the impression that currently insuperable technical or economic factors militate against rapid change in the energy sector, so that the public will continue to put up with the current state of affairs, despite appreciating the resultant acute dangers.

We must therefore make it our job to expose the real motives for the reluctance to take renewable energy on board – and already we are in the thick of conflict, because the interests of the energy complex are dressed up as seemingly objective constraints. Even pointing out highly subjective interests is characterized as 'lacking objectivity' and is treated as an attack. But as Karl Jaspers said, one must 'say what is, and do what must be'. Problems only become insoluble if even the possibility of a solution is subject to a conspiracy of silence.

So the electricity industry is virtually unopposed in its circumvention of new electricity market regulations requiring transparent separation between production, transmission and distribution. The long-overdue, hard-nosed response would be to put enforced functional separation on the political agenda, with the clear objective of placing at least the local gas and electricity distribution grids under municipal control, handing the medium-voltage grids over to the state, regional or county authorities, and bringing the national high-voltage backbone under the control of the national government, just as with the road network. The provision and maintenance of public infrastructure is no job for the private sector. When it comes to the privatization of public energy utilities, it ought to go without saying that only the power stations should be sold, and not the grids. This is the path that Italy has followed: privatization of the power stations, with the grids remaining in public ownership. This is a battle we must dare to fight.

It has only recently become public knowledge that German manufacturers of electricity cables have been colluding on prices for almost a century, creaming off billions of marks in the process. The case resulted in a fine of several hundred million marks. But what about the vastly exaggerated prices charged by the German electricity industry during the era of regional monopolies? It is becoming clear that that must have been the case, otherwise the newly privatized companies would not have been in a position to deliver instant price cuts of 20 per cent and more. The excess profits of municipal energy utilities were passed on to the municipal authority, and used to finance local services. The electricity companies, on the other hand, played Monopoly with the economy, building up capital reserves which now represent a strategic resource in the battle for the end-consumer against independent municipal utilities, the aim being to sweep both them and their more efficient local CHP capacity from the market. The current financial muscle of the electricity companies is evidence of the extent to which they have been leading the regulatory authorities up the garden path over recent decades, in order to secure approval for exaggerated electricity prices. Surely this is a multi-billion-mark injury to the electricity consumer, for which there must be a case to answer? A case of serious fraud, complete with conspiracy on the part of some regulatory bodies, and a matter of grave concern for state prosecutors and the courts? This is no water under the bridge, no more than it was with the cable industry, but a present-day battle — a battle that is not being fought because the grid and power companies regard themselves as a state within a state, and have licence to conduct themselves accordingly.

The EU competition authorities have imposed a fine on Daimler-Chrysler, a powerful car manufacturer, for operating a deliberate policy of charging less for its cars in certain markets in order to secure a greater market share — a well-known cartel trick. Yet in the electricity market, the power companies have been able to follow the same policy with impunity. They publicly admit to varying prices locally in order to undercut the competition as if it were the most natural thing in the world. Referring the case to the competition authorities and prohibiting such unfair competition is also a battle that politicians must dare to fight.

Tax-exempt reserves for nuclear waste disposal, which in the case of the German nuclear industry amount to €36 billion ($32 billion), equip the industry with a unique competitive advantage over all other firms. How and where the reserves are invested is entirely up to the companies concerned. Operators of nuclear power stations must clearly be obliged to put these reserves into a fund to be used solely for disposing of nuclear waste, and I have submitted a draft bill to this effect. However, no government has yet mustered the political courage to make such measures law.

The public, of whatever nationality, should no longer put up with the wilful burning of fossil fuels, and the damage to human health thereby caused, when it can be demonstrated that renewable energy is a real alternative. Nor should they put up with the continued manufacture of petrochemical products with all their damaging consequences, when a comparable product can be produced from biological resources at no extra cost. They should not tolerate diesel-powered motorboats whose unavoidable leaks have serious detrimental effects on water quality, when those boats could just as easily be run on vegetable oil, in which case leaks would simply provide good fish fodder. Following the Belgian egg scandal, swine fever and bovine spongiform encephalopathy (BSE), action is now being taken against the marketing of agricultural produce dangerous to health, with products being completely withdrawn from sale. And not before time! But why not also take action in the energy and chemicals industries? The right response would be to face up to the conflict, and withdraw fossil energy from the market everywhere where emission-free and non-toxic alternatives are available – for example, through a ban on sales of hydrocarbon lubricants and detergents, or a ban on the use of non-recyclable oil-based packaging materials. In Neckarsulm, Germany, a new housing estate has recently been constructed with a solar distributed heating system, and it functions smoothly and cost-effectively. In view of such exemplary proof that external environmental costs can be avoided, it is high time that comparable technology were made compulsory under building regulations everywhere.

With a litany of alternatives not taken up, although there is no further objective reason not to on grounds of quality, price or availability, there can be no more tolerance for those who flout the principle of tolerance in the face of society at large. If need be, lawsuits must be brought to take fossil fuel products from the market, just as happened time and again in the USA in the case of products damaging to the consumer. What the US consumer rights lawyer and presidential candidate Ralph Nader has achieved in court in numerous cases since the 1960s is also to be recommended in the energy and resource conflict. We should no longer brook the manufacture or regulatory approval of harmful products for which there are convincing and immediately accessible harmless alternatives of which the suppliers are aware. The damage done to humanity by non-solar energy supplies is greater than that caused by smoking, and the damage claims must be correspondingly higher. In the global struggle for resources, hard choices must be made for 'soft' resources. Whosoever ducks this fight has already lost.

10

Regionalization of the global economy through solar resources

THE CORRECT RESPONSE to the all-encompassing process of globalization, as most critics have realized, is to reconnect economic relationships with their regional basis. Most regionalization schemes attempt to either compensate for or work alongside the globalization process, or they consist of measures to make national economies more attractive locations from which to do business. Large – sometimes disproportionately so – quantities of public money are spent on such projects, including airports and facilities for trade fairs, without considering whether the new capacity is really needed. In many cases, it is regional economic policy designed to ensure global competitiveness that puts the regions at the mercy of globalization in the first place. If regionalization is to be an adequate response to globalization, it must take a completely different course: there must be a revitalized circular flow of goods and services at the local level, so that more activities can be taken out of increasingly global supply chains.

Just as vague as the concept of economic regionalization is the notion of what exactly constitutes a 'region'. The term usually refers to geographical areas delimited by state-defined boundaries. In the broadest sense, a region is an area that is small in comparison to its larger geographical and political context: a continent in relation to the planet, a country in relation to a continent, a borough in relation to a county or province. On a global scale, the EU and other subcontinental economic organizations and free trade areas are regional associations. However, regions delineated by administrative boundaries are too formal to embrace the opportunities that an ecologically oriented economy would offer for regionalization.

Proposals for stimulating the regional economy always run into the question of the extent to which regionalization policy may be 'protectionist' within the terms of the WTO and the European single market. If successful, policies that focus on attracting industrial firms do so at the expense of other locations. Although regional politicians have a direct responsibility to seize any opportunities that may arise, such policies do not resolve the contradictions inherent in the direction the global economy is taking. Policies that focus on filling the niches which have either been left by supraregional firms or are currently of no interest to them do make good sense, but they cannot counteract the general dependency on developments in the global market as a whole. The aim of regional policy must be to bring the focus of economic relationships from the global market back to regional markets. The crucial questions are therefore: how can this shift take place organically? How can regional economies be given a lasting structure that does not degenerate into a Sisyphean struggle against the global market, in which the efforts of today will be relocated or displaced tomorrow?

To anybody with a basic appreciation of natural ecosystems, it must be obvious that the flow of goods and services in the economy conflicts with the flow of nutrients and energy in the natural world. It is only really possible to make proper allowance for ecological loops at the regional level; everything else is a more or less imperfect approximation. As the laws of nature have priority over all market rules – and over any doctrine of economic planning – it is regional market links that we should be improving, rather than expanding global free trade. This must take place without recourse to the old tactics of economic isolationism, which have all too often been used to prop up unproductive structures or to place one economy at an advantage over another. The way the EU market has been closed to banana imports in order to benefit French overseas dependent territories is one such negative example.

The smaller the scale at which economic loops can be realized, the greater the chances for achieving ecologically sustainable economic activity. The supply chains are shorter, the middlemen can be cut out and it is possible to return resources directly to the local ecosystems whence they were extracted. That said, it does not necessarily make better environmental sense to organize economic activity on a regional rather than a global level in every case. Supplying the German market with solar power from North Africa rather than locally produced electricity from coal and nuclear power stations meets environmental criteria; likewise, shipping German-made solar panels to Nigeria makes more sense than burning oil from local reserves. Obviously, however, those same criteria would be even better filled if German solar power were locally generated, and solar panels installed in Nigeria were to come from local production.

A return to old national trade boundaries is neither desirable nor achievable at either the national or the global level. The idea that national economies should not erect trade barriers at will is in principle correct. But it is also correct that environmental scandal and social disaster would be the inevitable result of subjecting all instruments for safeguarding regional economic structures and environmentally sustainable practices to the same prohibition. It is in any case naive to believe that it is only backward, unproductive and therefore poorly performing sectors that are displaced by global agents. The victims of unfettered competition all too often include ultramodern and productive firms, because market access is controlled by the 'global players'. It is common knowledge that transnational corporations erect their own barriers to trade: statements to the effect that trade will be free if only the administrative barriers can be removed are absurd.

Neither undifferentiated regionalization nor undifferentiated globalization can yield long-term solutions. This begs the question: which economic activities should be fundamentally regional in scope, and for which activities is global free trade important? What generally applicable criteria and values make the case for regional markets? Supporters of globalization appeal to values like freedom, peace and the fight against nationalism, as if globalization were the modern expression of basic pacifist ideals. On the other side of the scales must be set social and environmental values. The standard demand is for social and environmental standards to be included within the remit of the WTO. Whether this can be achieved in a concise and consistent way, given the enormous variations in culture and levels of economic development from country to country, is, however, more than questionable. So what is to be done?

Regionalization effects through solar resources

The most important impulse towards regionalization would come from the transition to a solar resource basis. This is the lesson that the experience of conventional resource supply chains has to teach. The more thoroughgoing the transition to local renewable sources of energy, the stronger will be the regionalization effect that automatically results – right down to the smallest parish or ward. The process will take hold without any need for administrative boundaries, and the capital accumulated from energy cost savings remains within the local or regional economy. New and lasting jobs will result.

The extent to which renewable energy creates new jobs has not yet been quantified in any coherent way – ie, by comparing energy supply chains. It is probably possible to estimate the gross number of new jobs from renewable energy by deriving a figure for new jobs per unit of capital investment from comparable activity in other sectors. This method was employed by Wolfgang Palz at the EU Commission as part of the preliminary studies for the EU White Paper on renewable energy. According to his figures, trebling the contribution of renewable sources to EU energy supplies by 2010, from just under 7 to 20 per cent, would create two million new jobs, of which 800,000 would be in agriculture, 800,000 in the construction industry and the remaining 400,000 in the manufacture of technological equipment, solar technology services and consultancy.[1]

The real figure for additional new jobs will only be known once the number of jobs lost in the conventional energy sector during the transition has been subtracted. Calculating this net effect is far more difficult, as any serious study would need to take into account not only redundancies in power plants, refineries and traditional installation services, but also the entire workforce employed along the whole nuclear and fossil energy supply chain, from extraction of crude oil through to power station and pipeline construction. As 20 per cent production from renewable sources would leave the conventional energy supply chain largely intact, it can be assumed that initial job losses would be fairly small. Only conventional energy sales would fall, with concomitant increases in unit costs. But as soon as the demand for fossil energy fell to the point where no new contracts for extraction technology, power stations, replacement of ageing distribution infrastructure or conventional heating systems were being awarded, waves of redundancies would follow.

For this reason, it may be that the number of new jobs is lower in the long term than some optimistic estimates suggest. What is for certain, though, besides the creation of new industries on both the small and large scales, is that there would be considerably more employment in rural regions, in the construction industry, in the trades and in engineering consultancy, and that this would be widely and evenly distributed across all cities and regions. The new jobs would also be stable in the long term, as they would be tied to the locations of distributed energy production.

The manufacture of solar technology – solar cells, solar glass, fuel cells, wind turbines and small-scale hydro, Stirling engines, storage media, appliances with built-in solar panels, etc – will probably devolve to a few producers operating mass-production plants in just a few locations. The market for solar collectors and specialized PVs is more likely to develop a broader structure. Plant manufacture, however, will not bring as many new jobs as installation and maintenance services, or the forestry and agricultural enterprises that produce foodstuffs, energy and raw materials. Table 10.1 lists which renewable energy functions will be evenly distributed across regions, by comparison with the centralized organization that inevitably goes with nuclear and fossil supply structures. All energy functions, with the exception of the manufacture of generating plant, are overwhelmingly performed at local or regional level in the case of renewable energy, up to and including the financing of innumerable individual installations.

Table 10.1 *Regional distribution of economic activity: renewable and non-renewable resources compared*

	Heat and electricity production from renewable sources, with energy storage	Biomass for energy and raw materials	Nuclear power and fossil fuels
Extraction	None	Even	Uneven
Processing	None	Even	Uneven
Storage	Even	Even	Uneven
Distribution	Even	Even	Even
Installation of generating plant	Even	Even	Uneven
Operation of generating plant	Even	Even	Uneven
Maintenance of generating plant	Even	Even	Uneven
Energy supply model	Even	Even	Uneven
Local/regional tax revenues	Even	Even	Uneven
Regional provision of finance	Even	Even	Uneven

It is possible to bring local and regional economic agents into a nuclear and fossil energy supply, but this is more a chance occurrence and is not intrinsic to the system. Exploiting renewable energy, by comparison, results in a redistribution of labour from large firms and their geographical locations to regional or local situations and small and medium-sized undertakings, agricultural and forestry enterprises, and to tradesmen and the professions in the case of engineering design and installation services for renewable energy systems. Whereas, in decades past, jobs in municipal power plants were replaced by jobs in large power stations, now the reverse will be the case. Biomass farmers and foresters replace jobs in oil and gas extraction in Saudi Arabia and Russia, or in coal mining. Those currently employed in the lignite mines of eastern Germany could find new work in the same region, cultivating and harvesting biomass; power station installation engineers could move into the installation of solar systems; refinery workers could find new work in regional oil mills, biofuel production or in the processing of plant-derived materials.

Local councils and regional bodies with an independently managed budget and the power to raise taxes on commercial activity, and which also receive a proportionate share of the revenue from general taxation in their area, ought to have strong interests in seeing a swift transition to renewable energy. Local tax revenues would rise not only because money formerly spent on imported energy would remain within the local economy, but also through the new jobs that would result. Pure self-interest ought logically to push regional authorities into driving forwards the uptake of renewable energy on a large scale. Such investment for the future would pay for itself through the boost it would give to renewables businesses – quite apart from the accompanying environmental benefits, which really speak for themselves. By creating new jobs, the commercial exploitation of renewable energy also contributes more than any other conceivable initiative to achieving the original goal of regional economic policy, namely tackling social inequality.

Another impetus towards regionalization will come from the demise of the monthly energy bill and an end to the concentration of capital in the hands of the energy suppliers. Revenue from conventional energy supplies accrues to large public companies and their shareholders, in whose hands it also further fuels the consolidation and globalization activities of the business world, loosening their ties to national economies. Although the regionalization of energy supply induced by the switch to renewable energy will also lead to a loss of income in regions where existing conventional energy extraction and processing industries will be forced to close down, at the same time these regions would not be placed at a disproportionate or unacceptable disadvantage, as they will have the same opportunities for exploiting renewable energy as anybody else. Renewable energy levels out the international playing field and helps to deliver equality of opportunity, no matter where people live.

Large cities will also see improvements in their economic situation as businesses return and energy costs fall. New rural opportunities will put the brakes on rural depopulation, thus also lessening the pressure of migration on the cities. Urban–rural trade links will be strengthened as biomass production from farming and forestry takes centre stage. Urban demand for renewable energy or growing demand for ever more varied raw materials will spark the foundation of new rural businesses, ultimately leading to a decentralization of the national economy.

'Own implementation' versus 'joint implementation': opportunities for the developing world

Through the mechanisms already described, renewable energy contributes to a more equal distribution of income on a global scale wherever the process of replacing conventional energy is put in train. Renewable energy is the ideal tool for bridging the global gulf between rich and poor. Blindness to the societal consequences, and the enduring mythology and supply-chain influence of the conventional energy industry, are the only factors that can explain why renewable energy does not have pride of place in national strategies for economic development.

Developing countries whose currency is not yet freely convertible and which therefore have direct control over their foreign currency reserves are in an entirely different position. The obvious strategy here would be to reallocate foreign currency reserves to renewable energy by steering investment flows. As there is virtually no lead time between installing and commissioning renewable energy generation plant, developing countries could invest their foreign currency directly in imports of renewable energy technology. The investment would need to be costed over several years for a sensible comparison. The expected price of the amount of energy produced over ten years could be offset against the one-off capital cost of the generation plant: energy contracting on the scale of an entire economy. The calculations would almost certainly favour renewable energy. Developing countries thus have a real opportunity to make the transition from conventional to renewable energy under their own steam.

The majority of developing countries currently import both supplies of primary energy and the generation technology required. Domestic manufacture of the much less complex technology needed to exploit renewable energy, however, would yield an economic benefit that the industrialized countries have hitherto reserved for themselves, to the detriment of developing countries. From an economic perspective, it is irrelevant whether importers of renewable energy technology are required by law to manufacture their own equipment domestically, or whether the equipment is manufactured and sold by domestic firms. Either way, domestic production results in significantly lower costs as lower local wages mean considerable reductions in the large proportion of total cost due to labour. Countries with domestic renewables industries also gain the possibility of exporting plant to other countries – either to other developing countries which have not yet embarked on the same strategy (South–South trade), or as cheap imports to industrial countries, raising the woefully low level of South–North trade. In order to seize these opportunities, governments of developing countries must take the plunge and embark on the direct route to distributed energy systems for renewable energy without further ado. As already discussed, this is the only route they can sensibly take, because this is the only way to deliver high-quality energy supplies to rural communities, where there is desperate need for economic, social and cultural development capable of bringing forth viable enterprises in agriculture, the trades and other small businesses.

Development of the rural economy is also of vital importance for the production of regenerable raw materials, over and above the specific issues faced by developing countries. Without rural access to a modern energy supply, the production of bio-materials, for which developing countries have the greatest potential, will remain firmly in the hands of transnational agribusiness. This would deprive developing countries of the opportunity to produce raw materials for domestic consumption through a diverse range of small businesses and cooperatives. There would also be a clear danger that bio-materials would be cultivated in an environmentally damaging, extensive, export-oriented manner. An international bio-materials firm can go elsewhere once the land is exhausted: domestic cooperatives cannot, which means they are more strongly motivated to manage their farms or forests in a sustainable way.

This strategy is wholly impossible with conventional large-scale power stations, and not just because of the ongoing need for imports of primary energy. Importing a large-scale power plant can bring a developing country to its knees financially. As a proportion of total economic output, developing countries had already been far worse hit by the oil crises between 1973 and 1982 than the industrialized countries. Total debt ballooned six times over from around $200 billion to $1.2 trillion, and despite a number of debt relief programmes this debt millstone still hangs around their necks today. Further global oil price rises of comparable magnitude would drive many states into economic meltdown. Yet such price rises must be expected in the near future, between 2010 and 2020. As the next price leaps will not be caused by arbitrary decisions by the OPEC cartel, but rather by dwindling reserves, prices will remain high, and can only go up from there. Developing countries already run the risk of never being able to amortize costly new investment in conventional energy infrastructure.

Renewable energy thus represents the only, and at the same time a unique, opportunity for the economies of developing countries. New investment in conventional energy is a wanton use of economic resources. At best, such investment is only free of risk in countries that have their own fossil fuel reserves, which of course overlooks the environmental damage that results from developing countries' focus on fossil energy as a strategy for economic growth. The key, crucial point is that developing countries can follow the path outlined here under their own steam. Developing countries can no more enter a solar global economy by the power of development aid alone than industrial countries can make the transition from nuclear and fossil fuel to renewable energy solely through government grants and tax breaks. The main thrust of their efforts must come from within. To an even greater extent than for the richer nations, this is an issue that will directly decide the fate of whole developing country societies. And the path taken by the developing countries will also determine events in the wider world.

The result will be a historic irony if developing nations replicate the nascence of fossil fuel industrialization just as it enters its twilight years. There are sound, comprehensible reasons why today's industrial societies followed the path of technological development that they did. For today's developing countries, these reasons no longer obtain: the alternative is there for the taking! The way to satisfy the energy needs of the developing nations is not 'joint implementation', arm-in-long-arm with the industrialized countries and their energy corporations. Joint implementation mechanisms and tradable emissions certificates are the most frequently proposed suggestions for global action on climate change.[2] The underlying assumption is that because of the global effects of emissions from energy consumption, it does not matter exactly where emissions are reduced. It is therefore in the interests of all, the argument continues, that investment take place where it can have the greatest effect, and industrialized countries should therefore be permitted to meet their emissions reduction obligations under the international protocol on climate change through investments in developing countries as well as at home. This would also have the general advantage of transferring low-emission technology to developing countries.

The internationally-agreed concept of emissions certificates allocates an emissions quota to every country. In aggregate, the quotas should equate to the targets for global emissions reduction − up to 50 per cent by 2050, measured against 1990 levels. The allocated emissions quotas would have to be lower for industrialized countries and higher for developing countries than their respective current emissions, because of the much lower energy consumption in developing countries. Should an industrialized country desire to emit more than its allocated quota, it should be able to purchase or lease additional permits from developing countries. In theory, all countries therefore have an incentive to reduce emissions. Individual companies should also be able to participate in this trade in emissions.

Both mechanisms are designed to motivate governments to pursue a more rigorous policy on climate change. Both approaches have problems, on top of their failure in practice. Trade in emissions permits gives industrialized countries the option of buying their way out of the need to reshape their energy systems, despite the fact that it is the industrialized countries who most need to change. Developing country governments are also so mired in chronic budget deficits that it is highly questionable whether they would use the revenue from permit sales for purchasing and implementing energy-saving technology in order to limit their need for emissions permits. The joint implementation mechanism also illustrates the casual arrogance of the developed world, presupposing as it does that those guilty of causing climate change can put developing countries on the right path by transferring their energy technology. Here as nowhere else, following decades of profiting from the wrong technology, the industrialized world is skating on thin ice. Yet the real problem with both these mechanisms is that renewable energy has so far not featured prominently in discussions on combating climate change. The focus has been above all on lowering the consumption of fossil fuels, with almost everything hinging on emissions reduction rituals. Discussions on climate change prevention are generally regarded as impeding economic development, and thus meet with resistance, not least on the part of developing countries.

This problem would not arise if, rather than international obligations to reduce emissions, countries were to be set binding targets for the percentage of domestic energy generated from renewable sources, measured against current fossil fuel consumption. Such targets would automatically serve the objective of lowering greenhouse gas emissions, and the goal of increased energy efficiency also receives indirect consideration, because binding targets for renewable energy use are easier to achieve as total fossil fuel consumption falls. Focusing climate change agreements primarily on renewable energy use would also render them more acceptable to their signatories, as the shift away from the fossil resource base would no longer appear to be an economic burden. Renewable energy offers a unique opportunity – for developing countries above all, because they would be pursuing their own individual paths of development, rather than copying the mistakes of the industrialized world.

Regionalizing trade flows

Global trade in goods and services requires a global transport infrastructure. The faster the medium of transport, the wider the spectrum of goods and services sent round the world. The greater the transport capacity available, the lower the cost and the greater the volume transported. Faster and more economical transportation has probably done more for world trade than all the free trade agreements in the second half of the 20th century. Transport infrastructure has certainly made the current consolidation taking place in the global economy possible, allowing mass production to be extended to cover more and more markets.

The fossil energy industry has always profited from the growth in the global transport industry in at least two ways: through increased sales, and by expediting the industry's expansion into more and more walks of life, thus increasing sales still further. As this twofold sales boost lowers the energy industry's unit costs, it gains even greater scope for flooding existing markets with cheap products and thereby tightening its grip on them. Transportation networks powered by fossil fuels place a manifold burden on the environment. This is simply ignored by statistical success stories suggesting that economic growth has been decoupled from resource use in the transition to a 'weightless' economy. Replacing domestic production with imported goods only lowers energy consumption in the importing country; this is balanced by rises in the exporting country.

For this reason, among others, fuel duty exemption for international shipping and aviation is what has done the global environment the greatest damage. While the exemption for shipping is a hangover from earlier times, the exemption for aviation was only introduced after the second world war. The effect of these exemptions on the structure of the global economy began to be felt with the construction of giant freighters of over 100,000 tonnes displacement, and of large cargo planes.

The decision to accord tax-exempt status to aviation fuel was taken by the International Civil Aviation Organization (ICAO), an association of what were for a long time primarily state-owned airlines acting in their own interest. Although ICAO decisions are not binding, governments have always abided by them, in some cases even enacting new legislation. In Germany, tax exemption for aviation fuel is enshrined in the

Crude Oil Act; in the EU it is guaranteed by directive 92/81/EEC (ie, since 1992). These regulations are also anchored in innumerable bilateral agreements on aviation, of which Germany alone has signed over 120.

But the tax subsidies did not stop there. Their scope was expanded to include the purchase of aircraft, which is zero-rated for VAT and attracts further tax breaks, in Germany amounting to some 30 per cent of the purchase price. Airline operating capital held outside national frontiers is not liable to capital gains tax where the signatories to bilateral agreements so stipulate. Airports in most countries are not liable to property taxes. The shipping industry also benefits from tax breaks for the purchase of freighters. It has also become standard practice – without any serious political action ever being taken – to sail under so-called 'flags of convenience' such as that of Liberia, by which means shipping companies obtain yet further tax exemptions while also protecting themselves against most insurance and public liability provisions. The shipping and aviation industries have thus become a global tax-free zone.

These are not welcome topics for discussion. The German government's report on subsidies, for example, listed an annual subsidy – in the form of lost revenue – of only DM250 million (€128 million; $114 million) for aviation fuel and DM350 million (€179 million; $159 million) for shipping, DM600 million (€307 million; $273 million) in total. Only when a parliamentary question put down by the Green Party in 1995 insisted on the calculation of lost revenue in comparison to the standard rates of duty on petrol and diesel did the federal government have to admit a figure of DM8.1 billion (€4.1 billion; $3.7 billion) for the 1993 fiscal year. A statement to the parliament justified the discrepancy by claiming that the figures only referred to fuel consumed within national borders, as 'there can be no question of loss of tax revenue outside the tax domain'.[3] The same report listed DM35 million (€17.9 million; $15.9 million) of revenue lost on the purchase of ships and aircraft, a figure that is also obviously too low.

Criticism of these tax privileges has centred on the consequences for the environment of the growth in air travel. The tax breaks enjoyed by the shipping industry have generally escaped critical examination, with the result that many are either not aware of them or do not think them worthy of comment. The crucial economic consideration has been overlooked: these subsidies represent a subsidy for global trade which places it in a privileged position over local trade, confined as it is to road and rail. They are a targeted benefit to global firms, denied to firms with only regional scope. No attempt has yet been made to calculate the aggregate value of this subsidy. Its value to the global players, by comparison with average duty on fuels, is mostly likely to be around the $300 billion mark – to the detriment of regional trade flows.

This tax break thus does not just represent the greatest act of patronage in economic history, which, besides the shipping lines and airlines, primarily benefits exporters and the energy industry. It is also the largest single factor behind the accelerating degradation of the environment. Shipping and aviation account for around 15 per cent of all global oil consumption, and the proportion is set to increase in coming years. As emissions from aircraft cause at least three times as much damage to the atmosphere as ground-level emissions, aviation as a tax-exempt polluter causes around 30 per cent of the damage to the atmosphere. Oil-drenched international waters and numerous oil-coated coasts are ample testament to the environmental havoc wrought by the shipping industry.

Without these subsidies, global trade flows could never have reached their current scale. The agricultural sector above all would look very different. More than any other, the overwhelmingly US-based food-processing companies owe their rise on a global scale to international freight subsidies. Since the 1960s, the cost of shipping food products from the USA to Europe has fallen by 80 per cent.[4]

Tax exemption for freight has helped to destroy agricultural structures in developing and developed nations alike. By extension, it has also contributed to the appearance of slums; to the introduction of environmentally damaging cultivation practices in the pursuit of global competitiveness through enhanced yields; to the drain on government budgets from subsidies to prop up ailing domestic farms; to the falling quality of foodstuffs; to the global redistribution of nutrients through shipments of animal feed, which seriously compromise land quality in both exporting and importing countries; to the global dependence on a few, mostly US-based, suppliers of crop seed, thereby running the risk of acute famine should these companies be afflicted by, for example, drought or flooding; and to the poverty of modern food supplies and the demise of the agricultural sector, the indispensable basis of all societies.

Tax-exempt fuel for shipping and aviation has directly led to the deregionalization of economic links. It discriminates against regional suppliers. That it can now cost more to transport a comparable load from Passau to Bremen or from London to Glasgow than over the Atlantic or from Australia to Europe by ship or plane is a greater environmental problem than the – equally scandalous – preferential treatment of road over rail transport. Exemption from fuel duty privileges environmentally harmful over non-harmful freight transport, global over regional trade flows and industrial corporations over small- to medium-sized enterprises. It promotes the separation of product from consumer and the anonymity of the economic process, which runs counter to the purpose of a market economy. The rationale for artificially rescinding transport costs in this way is that there should be equality of opportunity among all producers, regardless of location.

Economic and environmental regionalization must respect and exploit the natural advantages of the right location. Subsidizing these advantages out of existence places a burden on nature and society that benefits distant, scarcely accountable economic agents. Freight subsidies are an attack on society and on the natural world. If the regions are to see a revival, then the subsidies for shipping and aviation fuel must be abolished at once. Freight costs that fully reflect the actual distance travelled will automatically lead to the regionalization of trade flows without any bureaucratic intervention. Firms will be motivated to manufacture in proximity to their markets, and thus to decentralize their production. Small and medium-sized firms will have better opportunities, as will domestic agriculture. The amount of energy consumed by shipping and warehousing will fall, lessening the strain on transport infrastructure. The scale of fossil energy supplies will be reduced, thereby weakening their position on the global market and hastening their replacement. It will be easier to market regionally produced biomass for energy and raw materials. Regionalized trade flows will be disintermediated, putting producers back in direct contact with consumers, with cost savings to both. The new relationships between producers and consumers will resemble the models described by the early 19th-century economist Johann Heinrich von Thünen in his work on the 'ideal city' and the 'ideal state':[5] the economic process as a series of concentric circles around the regional centres. These circles would not be administrative boundaries, but would result from varying production costs and distance-dependent transport costs.

Taking this step towards the regionalization of trade flows would probably have a wider effect than any other political proposal. It could also be immensely popular. Measures such as a tax on global capital transfers (the so-called Tobin tax) would be far more difficult to implement. Ending unfair treatment is a simple measure because it ought by rights to be a matter of course. It is precisely because such self-evident principles have been neglected for decades that the world is endangering its own survival.

The sustainable economy: global technology markets, regional commodity markets

That fuel-duty exemption for international shipping and aviation is already regarded as a matter of course and irreversibly shows just how one-sided the global economy is. The rules of the global market are drawn up to suit large, globally competitive companies. These corporate empires are in large measure a political creation. If the architects of the world trade regime had set their sights on the economy as a whole within its social context, rather than primarily following the interests of big business, they would have had to draft a framework that made it possible to hold companies to their social and environmental responsibilities.

Many 'modern' apologists for the global marketplace, however, denounce all attempts to widen the scope of market regulations to include social and environmental obligations as impracticable or as 'protectionism'. Protectionism is the new bogeyman of the neoliberal age, and yet all it means is security for individual livelihoods. The freedom or necessity for individuals to protect themselves is not normally open to debate. Anybody who decried the existence of police and military forces as 'protectionism' would be seen as a dangerous radical who would yield society up to the forces of naked aggression; at best there might be sympathy for this out-of-touch simpleton who naively trusts in the goodness of humanity, being ignorant of the ways of the world. The antiprotectionist dogma of a global free market has all the hallmarks of an ivory-tower delusion – to think that economic life of all things, that daily battle to survive, would be devoid of aggression! This naive view of the world is proclaimed primarily by the most successful practitioners of economic aggression, whose talk of markets and equality of opportunity is no more than self-interest. There is no question that protection is an essential ingredient of economic life. Those who claim that competition always favours those with the best product and the greatest productivity are preaching sanctimonious economic parables. The world hangs in the balance when ivory-tower dogma becomes global economic practice. Every civilized society needs defensive mechanisms, and that applies also – and especially – to its economic system. The question is simply what society seeks to achieve and protect. Is it the mechanism best suited to satisfying human material needs and to productive economic activity, and is it based on objective and generally applicable principles, or merely the selfish interests of corporations or states?

Whether social and environmental standards should be incorporated into the treaties regulating world trade (GATT, the General Agreement on Trade in Services (GATS) and the Agreement on Trade-Related Aspects of Intellectual Property Rights (TRIPS)) goes to the heart of the debate on protection in the global economy. Such standards would allow countries to close their borders to, or impose additional tariffs on, goods whose low price evidently relies on worker exploitation and environmentally damaging production processes. It is within the bounds of possibility that lengthy and tortuous negotiations might result in the incorporation of social and environmental standards into an expanded global trade treaty. But it is unlikely that such standards would be sufficiently tightly worded to bring about socially and environmentally sustainable practices on a global scale, or that states would have sufficiently powerful measures at their disposal to be able to mount an effective defence of sustainable business forms.

Many people see GATT as the Bible for the global economic system, a sort of world constitution that has primacy over all other treaties – or at least over those with direct implications for business activity. But this is ideology talking. In reality, GATT, GATS and TRIPS are just three treaties among many, including the conventions of the International Labour Organization (ILO), which, among other things, guarantee the right to join a trade union and the right to negotiate salary and conditions of employment in every signatory state; or the UN Convention on the Law of the Sea, the Antarctic Treaty, or the Convention on Biological Diversity and the Convention on Climate Change. All these treaties have equal standing in international law, which recognizes no umbrella treaties other than the United Nations Charter and the UN Convention on Human Rights, both of which have constitutional character or should be interpreted as such. In taking it upon itself to put free markets before existing social and environmental agreements, the WTO is exceeding its mandate.

Demands for the inclusion of social and environmental standards within the remit of the WTO would indirectly accord the organization a role as an international arbiter in economic, social and environmental issues. The remit of the WTO, however, must be bounded by other international agreements. Cases of conflict between the WTO and the ILO conventions or a global agreement on the environment would be referred not to the WTO, but to an international court. It is because the WTO is accorded greater authority than provided for in international law that no such conflict has yet been tried in a court of law.

In sum: there have long been provisions for social and environmental standards in international company law, as Julius K Nyerere has pointed out. Prior to his death in October 1999, the former President of Tanzania and Chairman of the South Centre was firmly opposed to the inclusion of social standards in the WTO treaties.[6] Whether future trade agreements explicitly acknowledge the validity of social and environmental agreements is irrelevant: they have force in international law. Attempts to incorporate them into GATT and GATS put them at the mercy of negotiations on the future development of the WTO, when WTO delegations have no business deciding which other international agreements shall have force in international law. Heribert Prantl from the newspaper *Süddeutsche Zeitung* neatly summarizes the overblown aspirations of economic liberalization in an ironic reversal of the German constitution, the Grundgesetz or Basic Law, whose first and second articles declare the freedom of the individual to be inviolable: 'German competitiveness is inviolable. The undisturbed practice of investment is guaranteed. No-one may be compelled against his conscience to protect the environment, to respect privacy, to provide protection from summary dismissal or to enact any other measures that may place restrictions upon him.'[7] International law deserves equal protection from erosion when it conflicts with WTO rules.

Which rules and rights are respected, especially in company law, and even more particularly in disputes between countries, also depends on whether those affected have the courage to stand up for their rights. Without trade unions, consumer organizations and a few bold individuals, many national laws against the ruthless application of economic power would exist only on paper. Governments only decide to bring suits against other countries under international law after carefully weighing up the implications for their relations with the country in question. If international environmental law is to be enforced, then individuals must be allowed to bring suits before international courts, and not just governments. There also needs to be an international court of environmental law for this purpose, parallel to the International Court of Human Rights.[8]

Whether countries would abide by the judgements of such a court remains a question of international power politics. The USA takes the liberty of erecting trade barriers even in the face of WTO rulings – for example, against countries that don't observe the trade embargo against Cuba. As no international organization has sufficient force at its disposal to ensure that rulings by the WTO, ILO or international courts are respected, compliance is always a question of the balance of power between those involved. A verdict is more likely to be respected when it favours an economic power with the ability to impose its own sanctions, whereas rulings in favour of those without further influence are often empty words if the more powerful party chooses to ignore them. While international law may impede the powerful in the pursuit of their own interest, it rarely stops them entirely. As national economies become ever more dependent on transnational corporations and the global market, even for basic supplies such as food, energy and raw materials, they are increasingly at the mercy of an economic force against which no recourse to international rules on the environment, social issues or even trade can help. Whether right is on their side or not, countries can be blackmailed with threats to cut off essential supplies if they do not yield to the economic powers who control the world's resources. Hence resource autonomy at the level of the national economy is a fundamental objective that grows in importance as economic relationships become more international in scope. From an unblinkered perspective, this is in the economic interest not just of small nations, but of all countries, including those with the highest consumption of resources, the USA above all.

Already, even leading industrialized countries are harming themselves if their governments:

- act in the interest of the global market power of 'their' international resource companies, whose products destroy both the global and their own environment;
- provide military forces, including 'rapid reaction forces', to this end; and/or
- take the flak for the global economic plundering of resource reserves by the global players.

Governments that act in this way take on the role in the public consciousness of the 'ugly American', as Senator Fulbright noted back in the 1960s, or of the 'ugly European' or 'ugly Japanese', standing in for a few firms with US, European or Japanese names. Even the governments of the industrialized world are faced with the question of whether they would not better serve their interests by embarking on the road to a solar resource base, rather than continuing to defend the 'free' global commodities market.

Free trade should only continue to hold sway in the market in which the two core economic arguments for it actually obtain: that is, the market for technological goods. The first argument is that protectionist isolation is ultimately damaging to a manufacturing economy, as there is less incentive to seek productivity gains in a closed economy. The longer insulating tariffs remain in force, the further behind the economy falls as its productivity drops. The second reason is that every product for which there is a demand should be available everywhere, even if it is not domestically produced.

The flaw in the reasoning occurs where global market rules are extended to cover those products whose production depends on factors not under human control. In respect of commodities, this is always the case. Denying this difference and subjecting energy, raw materials and foodstuffs to the same global market rules as manufactured products is an error that criminally disregards the laws of nature. It is an error of the same magnitude to extend the free-market principles applicable to manufactured goods to cultural goods, the spiritual and intellectual resources of humanity and all its diversity in language and way of life.

Cultural levelling and the abuse of the natural environment, through to its irreversible depletion and destruction, are the consequences of these errors. Globalized markets have led in practice to an economic structure that rides roughshod over all moral principles and has no regard for the natural basis of life on Earth. The market principle is exposed as extremist economic dogma. A sustainable market regime must rescind all global market regulations pertaining to the resource industry, restricting their scope to manufactured goods. The global market for technology and technological products must be truly free and open to all comers. In the commodities market, however, every national economy must have the ability to prioritize locally produced over imported resources. This means:

- Domestic agricultural commodities such as grain, milk, meat and vegetables must have priority. Anything that cannot be produced domestically – in northwest Europe, this includes tropical fruit and vegetables and olive oil – must be freely tradable. The same applies to all foodstuffs where demand cannot be satisfied from domestic production. Equally, there can be no artificial restrictions for quality goods such as wine and speciality cheeses. This basic pattern can be expanded upon with regional preference regimes for basic agricultural commodities in countries with large land areas or in trading blocs such as the EU or the North American Free Trade Area (NAFTA).
- Domestically harvested or extracted energy and raw materials must be given priority over imports. This would automatically boost the market for renewable energy. Even countries with conventional energy reserves would lose their current cost advantages with the end of mass production for the global market.

This approach would be considerably better targeted, less bureaucratic and of more direct relevance than defining global standards for the entire domain of social and environmental policy. To be effective, standards necessarily have to be detailed – an enormously complicated undertaking, given the very different economic and environmental starting points in each country and their very different social and environmental legislation. The approach I propose returns the economic initiative to the level of national and regional economies, vital if social standards are to be upheld. By starting with the resource base and according priority to regional resource flows, it incentivizes sustainable business practices far more than would an environmental policy that merely sought to rein in the environmental excesses of traditional business through a plethora of individual regulations. How can a system that barely works on a national level be made to function on a global scale? Traditional environmental policy expends considerable administrative effort on improving conventional resource use. A modern environmental policy must focus on environmentally sustainable resources and traffic reduction. A global drive towards regionalized commodity markets furthers both these aims, being the swiftest route to a self-regulating sustainable economy. Regional markets make it possible to largely circumvent the intermediate resale agents that are a particular characteristic of the global market in agricultural products, and thus to avoid higher prices for consumers. It is the middlemen who take the money from the pockets of agricultural businesses, thereby accelerating their demise. Producer-organized direct marketing for foodstuffs and regenerable resources can, with political support, remove the need for intermediate trade, including the use of labels to indicate the origin of the product. One common argument against such a thorough regionalization is that developing country and East European agricultural and resource producers would lose their export markets. Yet all these countries have difficulty feeding their own populations, and in most cases it is not the population at large or domestic companies that profit from resource exports, but rather transnational resource corporations.

Trade not talk: beyond the energy industry

'The congress dances.' Such was the disparaging view taken of the 1815 Congress of Vienna, where for many months the diplomatic envoys of the governments of Europe negotiated the political future of the continent, following the 'big bang' of the French Revolution (1789) and the Napoleonic Wars. In terms of results, however, that congress achieved more that the negotiations, now in their tenth year, on the international treaties supposed to establish a global environmental protection regime. An unprecedented conference marathon produces agreements, when it produces anything at all, that are promptly ignored. Worse still: elsewhere, at the same time, the same governments are signing international agreements whose direct consequence is to negate the agreed environmental targets, from the WTO treaty to the European Energy Charter through to agreements on aviation. Unlike environmental accords, these agreements are usually observed. And yet there seems to be no alternative to global governance on environmental issues. Hence the many conferences, preparatory conferences and follow-up conferences at which protective measures for the climate, endangered species, land, sea, tropical forests and the ozone layer are negotiated. Non-governmental organizations (NGOs) also concentrate their activities largely on shadowing the global conferences in order to present them with their demands. The result is a growing body of international environmental law,[9] whose further development makes undoubted sense. But the fate of the planet must not be made contingent on the success of efforts in such conferences, be that explicit environmental treaties or new WTO rules. Calls for and attempts to initiate international agreements present an ideal excuse for governments not to undertake even the most urgent revision of policy on global resources, allegedly because an international understanding is a vital prerequisite – usually without any particular effort to bring such an understanding about. 'Talk globally – procrastinate nationally', was how my book *A Solar Manifesto* characterized this abuse of international negotiations as a smokescreen for business as usual.[10] Ross Gelbspan's book *The Heat is On: the high-stakes battle over earth's threatened climate* describes in detail how the ambitious UN Conference on Environment and Development in 1992 in Rio de Janeiro, at which Agenda 21 was agreed, has remained 'bogged down in diplomacy'.[11] Even with dedication on all sides, the broad consensus needed to negotiate an international treaty that will be ratified by a large enough number of countries to give it the status of international law makes for a tortuous process. Not only do the participating countries make concessions to each other, they also grant concessions to transnational firms. Serious and increasingly acute dangers cannot be resolved by such long-winded and ineffectual methods.

It is noteworthy that the only international agreements on the environment to have been achieved so far do not significantly touch on the interests of big business. The Antarctic Treaty, the Treaty on Maritime Law and the Montreal Protocol on the use of ozone-damaging gases are all examples of limits placed on environmentally damaging activities that have either not yet begun or – in the case of the Montreal Protocol – did not endanger powerful economic structures. Protracted efforts to achieve international agreement are scarcely amenable to fast-tracking. There are several proposals for global action, including the proposals contained in *A Solar Manifesto* for an International Renewable Energy Agency (IRENA), parallel to the International Atomic Energy Agency (IAEA), to organize international non-commercial technology transfers for renewable energy on global scale; and for a 'proliferation treaty' on renewable energy in the form of a protocol appended to the existing non-proliferation treaty on nuclear technology. But as useful as these initiatives may be if they succeed, it would be negligent to rely on successful negotiation alone.

In the case of renewable resources, it is also simply inappropriate. There is no need to agree access arrangements for a resource that is universally available. The only requirement is the right technology, for which no international treaty is required. Ultimately, it is a question of individual action, the only possible obstacle being national, European and international market rules that directly or indirectly favour the fossil resource industry. This absurdity must be brought to an immediate end, without waiting for changes to international treaties to establish market precedence for self-sufficient and sustainable resource use. International and European law already enshrines a sufficient number of such principles so that it would be possible merely to give these existing principles priority over market rules. No country need allow itself to be compelled by market rules to accept imports of crude oil or coal or electricity from nuclear or coal-fired power stations to the detriment of renewable energy. Every government can give priority to renewable resources, even if that means taxing environmentally damaging goods. It then has only to apply the same policy domestically, and give priority to renewable over fossil resources in the domestic economy. Governments have greater freedom of action than is generally assumed or claimed: it is just a matter of seizing the opportunities that already exist. And because the transition to renewable resources is not a burden but brings important advantages, there are no real economic constraints. The constraints are mostly imaginary.

Consensus among the masters of the existing superannuated global resource industry is not a precondition for the transition to a solar resource base. As long as this misconception holds sway, the opportunities that renewable energy and resources offer will fail to be adequately and appropriately exploited. We must be prepared to think and act outside the energy industry box. The solar economy will not blossom in debating chambers and boardrooms, but among practitioners of sustainable architecture, cultivators of energy and materials crops and designers of energy technology. Action must be taken locally, not globally.

11

The visible hand of the sun: blueprint for a solar world

SOCIAL DEVELOPMENTS MAY seem unpredictable, but that does not mean that the course they take is wholly determined by chance. Provided that no great war or natural disaster plunges everything into chaos, they follow a clearly discernible pattern of events. The fossil energy system has shaped the global economy from its inception, leading the world to the lip of the abyss. A global shift towards renewable resources will overturn the structure of the fossil global economy and draw mankind back from the abyss, towards a sustainable future. Sooner or later there will be general recognition of the need for fundamental change.

When the way to a lasting supply of clean energy becomes clear, people will not let the opportunity slip through their fingers. It could be a long time before the optimum path is known, and the route chosen will determine the pace of change. But the capacity of existing norms and structures to hold up the transition that the planet and its people so badly need may mean that it comes too late.

Over the course of history, many civilizations have fallen victim to their failure to wake up to mortal dangers. The fossil fuel crisis places the entire world in such a life-or-death predicament. Political and business leaders have so far shown an inability to rise to the challenge. They shrug their shoulders, content to blame anonymous market forces, to the approval of all those who hold that political institutions are in any case terminally sclerotic. There is some truth in this, as today's politicians seem to have no stomach for decisions that run counter to established business interests. Never before has political failure been so comfortable. This, too, we owe to the market – for now.

Political initiatives can and must drive and accelerate the replacement of fossil resources. Waiting for reserves to be exhausted is not an option. Their primary task must be to end privileged consumption of fossil fuels – which means abolishing direct and indirect subsidies and absurd tax exemptions – and to blaze the trail for renewable resources. If renewables can achieve rapid market penetration and revolutionize energy consumption, then there is hope for the world even in the declining years of the Industrial Revolution.

History tells us that the Industrial Revolution did not take place everywhere at the same time or even to the same extent, and that it was anything but harmonious. It was not the result of any political plan; it took off because at the time, the new technologies offered by far the best use of resources and thus the greatest potential for economic development. For this reason, it became the pre-eminent model for economic development. But the number of losers is growing exponentially as the disastrous environmental and social consequences become ever more acute, and greater numbers of the former winners now figure among the losers.

Whether the world can make the transition from fossil and nuclear energy to renewable energy in time will ultimately determine the historical status of the Industrial Revolution: a new era of opportunity, or the first step on the road to doom. Idealism alone will not bring about a solar technological revolution in energy supplies. We must recognize and exploit the economic potential of solar resources, and have no truck with the unscrupulous motives of nuclear and fossil energy companies. The road to the solar global economy is like a stream that takes a straight or winding course depending on the topographical hurdles to be overcome, flowing faster or slower depending on the size and speed of its tributaries, and finally swelling to a river with the power to reshape the landscape that surrounds it.

The first priority of every society must be to secure essential supplies of water, energy, raw materials and food. This truth is only overlooked in societies (and in the scientific community) where easy access to necessary resources has for a long time been a matter of course. This era of plenty is coming to an end. If a source of vital supplies dries up or becomes polluted, then people are forced to move to a new source of supplies or make a concerted effort to find a replacement. For this reason alone, merely moving from a recently exhausted source of supplies to another on the edge of exhaustion can only provide a temporary respite. The seemingly boundless capacity of the global market still leads people to rely on switching at will from one fossil fuel to another, although the limits to global supplies of these resources are clear to see. Companies are now beginning to see a major business opportunity in the area of water supplies, precisely because many regions are already feeling the pinch or can at least see the limits to their supplies. In view of the disastrous practices of the expanding agribusiness industry, anyone who relies on a boundless global market for food supplies is clearly living in cloud-cuckoo-land. The widespread belief that global transport capacities mean that dependence on anonymous suppliers for the basic needs of society is no longer a problem just goes to show how thoroughly the world has been led up the garden path. Whether they choose to see it or not, societies everywhere must focus their efforts on the use of environmentally sustainable and inexhaustible resources – and in particular on those that require the least effort to extract while offering the greatest economic benefit. The laws of economics compel us to seek lower real costs and hence to return to supplying humanity's essential needs from primarily local sources.

It is in the area of energy supplies that humanity has moved furthest from the natural world that sustains us, and it is thus in this area that past errors will be the most difficult to rectify. Hence our top priority must be a realignment towards renewable energy. In agriculture, it is only in recent decades, albeit at an accelerating rate, that ever-greater swathes of humanity have become estranged from local supplies. This has become the second crucial issue that we face, as sustainable long-term food supplies come under increasing threat. Once fossil and nuclear have been replaced with renewable energy, and regional agriculture has undergone its crucially necessary revitalization, a broad shift towards renewable resources will also follow.

In a solar global economy, water supplies will be secure in the long term; farm- and woodland will be managed sustainably; the essential needs of humanity for energy and raw materials will be met from inexhaustible sources; and energy will be supplied almost exclusively – and food and raw materials to a far greater extent than hitherto – from regional sources. The solar global economy is the economically and environmentally superior model, and it is the world's greatest social and cultural opportunity.

It is the task of the modern, environmentally conscious age to force this transition to a solar global economy, thereby overcoming the doomed fossil industrial age that has not only closed its eyes to the life-and-death choices that confront it, but utterly denies that such choices exist. The US philosopher Arran Gare writes in his book *Postmodernism and the Environmental Crisis* that disorientation 'has been made a virtue, and the absence of fixed reference points is celebrated.'[1] He portrays a generation mistrustful of the wider picture and of large-scale solutions. Such crises of identity and the loss of confidence in the future of society have always made an appearance when the existing social model has lost its credibility. But that is no excuse to abandon all convincing models.

How can we believe that there can be no more grand designs, no convictions and no more strength at a point when the global environmental crisis threatens to bring mega-catastrophes of untold scale? Such ideas speak of intellectual and moral poverty. Why is it that even social democratic parties, which owe their origins to a faith in human progress, and even green parties are showing an unmistakable preference for other topics than environmental planning for the future? Why does the modern age so lack courage and conviction? Why is this capitulation celebrated as 'facing up to reality', and how is it possible for cloud-cuckoo delusions to pass themselves off as neorealism without being laughed out of court? Disorientation and the lack of a moral compass are the symptoms of an age in which the multitudes who are embedded within the current system have utterly lost the power to imagine an alternative. That goes particularly for the many people who have recently had their illusions shattered by the failure of truly unrealistic utopian ideals, ideals to which they clung by censoring out unwanted truths. We are left with people who speak vaguely of renewal, but whose real aim is to preserve the comforts that the present still affords, at least for themselves, all in the knowledge and acceptance that that is something that ever-decreasing numbers of people can aspire to.

Forwards – towards the primary economy

Bringing about the vitally necessary displacement of fossil by renewable energy and resources will open up a future that gives a new impetus to the primary productive economy. No longer the economic leftovers, agriculture and forestry will become the new and lasting motors of the economy as a whole; not picture-postcard nostalgia, but modern, forward-looking enterprise; not a declining industry, but a major source of new employment.

The industrial and, to an even greater extent, the post-industrial age saw the classical primary sector of the national economy – most notably agriculture – being marginalized as the secondary manufacturing and tertiary service sectors expanded. Employment statistics showing the ever-declining proportion of the workforce employed in the agriculture and forestry sector can, however, lead to erroneous analyses and correspondingly erroneous conclusions. The current primary sector actually employs considerably more workers than the statistics would suggest. This includes all those employed in industries further up or down the agricultural supply chain, who would formerly have been directly employed by agricul-tural enterprises: in the production of fertilizers and pesticides, seeds and animal feeds, in energy supply and in marketing. Then there is employment in shipping agricultural inputs and products, food processing and the manufacture of agricultural machinery. None of these occupations find their way into statistics on the agricultural sector, although without it none of them would exist.

The received wisdom of the industrial and post-industrial modern age is that this development is irreversible. This view, common as it is, demonstrates both prejudice and a lack of imagination. If a country has to import technological products or services on a large scale, nobody would dream of concluding that the industries concerned have no future. The response is rather to attempt to re-establish them on domestic soil. Yet when agricultural production moves abroad, it is thought to be gone for good. Even dire news of global environmental trends cannot shake these negative attitudes to agriculture, although the obvious and swift consequence may be that the global trade in agricultural produce is no guarantee of stable food supplies.

If all those who would naively place all economic activity in the hands of the global market only understood the immutable laws of nature that override all ideology and dogma, then they could not fail to see that, while technology can be globalized, in the long term, resource supplies cannot. Manufacturing plant and services that do not require direct personal contact can be expanded and relocated almost at will. Croplands and other natural resources cannot be so easily moved. Agricultural productivity is not simply a function of education, efficient organization of labour and optimal use of machinery. It also depends heavily on the invariables of the local geographical and climatic conditions. This is the crucial difference between the primary and all other sectors, which the undifferentiated ideology of the global market simply ignores. If the current state of affairs continues, innumerable countries risk losing their agricultural sector as global production concentrates on the most geographically and climatically favourable arable land, the fertility of which will then be lost all the more quickly to overproduction.

The assumption that agriculture must develop business structures equipped to meet global market conditions, following existing developmental trends, is also criminally negligent in its acceptance of unhealthy consequences for society, especially in the countries of the developing world. If the greater part of the world's 600 million farmers and their families, 3 billion people in total, were to make their way to

the cities, leaving the land to be worked by agrifactories producing ever fewer products, the economic and cultural
consequences would be incalculable. This model of economic
development is jeopardizing our future. Much is made of the
knowledge economy as a strategy for dealing with the individual and social challenges to come. What most people fail to
realize, however, is that we will be in no position to face these
challenges if the existing practical knowledge of land management and plant husbandry of hundreds of millions of farmers
were to vanish along with their livelihoods.

The future of society can no longer be secured if the
economy is not structured around primary production.
Reintegrating primary production back into the national and
regional economy is of paramount importance. It will become
indispensable as renewables begin to replace fossil resources.
Economic realignment towards renewable resources and the
biotechnology that drives their use will give further impetus
to this process. As agriculture transforms itself into an
integrated food, energy and resources business, it will start to
grow rather than continue to shrink.

The real business opportunity for the agricultural sector
lies in this cross-sectoral synergy. This, together with the
concomitant ability to produce fertilizers and pest deterrents
on site, is what will liberate farming from its suppliers, the
chemicals and energy industries. It will also bring jobs back
out into the countryside, partly with new roles, but also with
wholly new opportunities. The person specification for a
farmer capable of dealing with the whole spectrum of plant life
is as demanding as any: the adaptability to learn the land,
climate and nutrient requirements of a variety of plants; and a
solid grounding in biology and biochemistry, and in the latest
harvesting technology. In a solar global economy there will be
a need for more independent farmers and more farm enterprises; agriculture will once again offer secure jobs to many
more people. Agriculture will also deliver a huge number of
less demanding jobs, of the kind that the service economy is
crying out for. But as long as driving a JCB in an open-cast
lignite mine, assembly-line work or the new domestic service
jobs are seen as more valuable than the comparatively demand-

ing and varied work of sowing seed, operating harvesting machinery, managing woodland or operating desiccation or biogas plants, the modern world with its cultural blinkers will fail to see this future.

This model of a rediscovered primary productive economy has several important implications, which show the agricultural trends of recent decades, along with agricultural policy and the current direction of the biotech industry, to be thoroughly short-sighted:

- There must be an immediate stop to the continued haemorrhage in the farming industry. Otherwise we will have to painfully rebuild in the near future what is now carelessly being sacrificed to the market. That is not an argument for maintaining the current subsidies indefinitely. Farming could be more effectively supported with help to organize the independent or communal production of energy and nutrients and to establish structures for regional direct marketing, such that through reduced costs and greater profitability farmers can once again support themselves. Subsidies should come with a 'sunset clause', and be redirected to help finance the installation of bio-energy processing plant and the conversion of agricultural machinery to run on bio-fuels. There must be support for regional marketing cooperatives for food, energy and solar resources.
- Seed supplies must be secure and freely accessible. No country will be able to manage without national seed banks providing immediate and universal access to the whole variety of plants likely to be demanded in future. Plant seed is a cultural treasure-chest that belongs to the nation, and seed stores from which seed can be purchased must be an integral part of the public infrastructure of the agriculture of the future, thus providing agricultural businesses with as level a playing field as possible.

If these opportunities are not to be squandered before they can be properly seized, there must be political action to call a halt to the recent growing wave of patents on genes and gene sequences. The natural fruits of evolution must remain in the public domain, freely available to all farmers. At most, only processing techniques should be patentable. It is in the vital interest of every country to put a stop to the patenting of genes, not just for reasons of human and bio-ethics, but also for the future development of the national economy. There must be an end to this most heinous case of dispossession in history, the dispossession of society by private companies. Only a ban on gene patents can prevent the rich potential of biological resources coming under the control of a few global firms before that potential has been realized on a significant scale. Ceding control of biological resources to biotech companies accords them a level of global power that all the global and colonial empires in history cannot rival. It is essential that no effort be spared to break this power if there is to be a renaissance of the primary productive economy.

Work and the solar economy

In his book *Die energetischen Grundlagen der Kulturwissenschaften* (The Energy Basis of the Humanities), published at the beginning of the 20th century, the winner of the Nobel Prize for chemistry and expert in the sociology of energy systems, Wilhelm Ostwald, characterized energy as 'everything that comes from or can be converted back into work'.[2] This understanding of energy covers three different types of work: human labour, native energy and the work done by mechanical devices and machinery. The question of the 'end of work'[3] must therefore also be seen in the context of energy and resource structures. The transition to solar resources will have effects on the future of work as a social institution over and above the new jobs it creates.

In the industrialized countries, machines have replaced human labour. For a long time workers saw no benefits from this, as the technology was primarily used to increase productivity, not to make the remaining jobs easier for people to do. The burden on workers was only ever lightened as the result of action by politicians or trade unions. The future will be no different: technology may make improvements in living standards possible, but this is not guaranteed to happen when others have other aims in mind. New computer and information technologies make it possible as never before for people to let motors and machinery do the work for them; mental and technical skills are replacing physical skills faster and more comprehensively than ever before.

This brings a new edge to the dispute over how the product of labour should be distributed across society so that everybody can make a living for themselves. If everybody has a sufficiently well-paid job, then the product of labour is distributed through employment incomes – with all the squabbling over fair and adequate pay that inevitably ensues. If there is not enough work to go around or growing numbers of people are excluded from the direct redistribution of income, then income must be redistributed across society through shorter working weeks and/or state minimum income guarantees. This is the new big issue for social policy. At the same time, there is the increasingly interesting question of how people spend their time outside work. Mathias Greffrath talks of a 'three-shift society', in which people spend a third of their time in paid work, a third in unpaid voluntary activities and a third on their own needs.[4] Johano Strasser emphasizes the indisputable need for a redistribution of work in his book *Geht der Arbeitsgesellschaft die Arbeit aus?* (When the Working Society Runs Out of Work).[5]

As long as this redistribution can be only imperfectly realized, the question of how people with no income from paid work can make a living will be the key issue facing any community. In theory, the output from energy and technology is so great that ever-decreasing numbers of workers are required. But the question that becomes more urgent by the day is how political institutions can siphon off the necessary income from highly productive firms, which are increasingly organized on a transnational basis and which effectively operate beyond the borders of all national governments. It is sheer fantasy to suppose that international political institutions could impose taxes on excess profits to redistribute on an individual basis. This makes it all the more necessary to take a close look at the energy component of work.

Industrial society has forgotten that the sun is the greatest and most versatile source of energy available to life on Earth, and that the power of the sun can be used to save human labour. This is equally true of agriculture and forestry. Here, too, human labour has been replaced by machinery and fossil energy. If the work done by fossil energy were to be done by the sun, then agriculture would reach the quintessence of its actual economic potential. The same can be said of work in non-agricultural sectors. The Industrial Revolution was reduced to replacing human with mechanical labour while also increasing the work done by fossil energy. Replacing fossil fuel work with solar technology will radically reshape the work society. With the sun doing the work formerly performed by fossil fuels, the total cost of work to society is reduced to the cost of human labour and technology.

Leveraging the sun will, as has been described, result in a more equitable distribution of jobs across the regions. The concomitant regionalization of the economy will make it easier for governments to use taxes to finance public services. Individual living costs will be permanently lowered, thereby making it easier to find a solution to the big question of how to provide everybody with the opportunity to live a dignified life free of destitution, as the cost of state income guarantees falls. Renewable energy helps us to escape the environmental catastrophes of the fossil industrial age, and brings with it lower energy costs. At the same time, it also reduces the risk of damage to human health – the costs of which are borne not by the perpetrators but by society as a whole – and thereby reduces the cost of maintaining a health service. Without the shift to renewable energy, disaster prevention and disaster relief will consume ever-greater proportions of public and private funds, until increasingly frequent catastrophes overload our capacity to cope, and social and civil order can no longer be maintained by even the best-equipped security forces. Society also foots the bill for the current bout of spending on military equipment to ensure the security of the remaining fossil fuel reserves, and the public is led to believe that this is in their interest.

Even if the proportion of work done by people continues to fall in comparison to the proportion done by machines and by the sun, society will face this issue from a wholly different starting point than today. There will be less potential for aggression, in part because the environmental dangers that now threaten us will have been averted. The only remaining question is how people spend their free time – and the answer will come not from legislation, not from market rules, and not from energy systems or technology, but rather from traditions, cultural norms, the human capacity for education and social interaction, and from the cultural achievements of society.

From the bounty of the sun to global economic prosperity

The Earth is rich, and it owes its wealth to the sun. That this wealth is today more often burnt than used and preserved for the future is the greatest economic nonsense imaginable. And then to call this destruction of resources 'economic growth' makes a mockery of the phrase. This is not economic growth, but economic destruction, and it leads not to Adam Smith's 'wealth of nations', but rather to Elmar Altvater's 'poverty of nations'.[6]

The fundamental problem with today's global economy is not globalization per se, but that this globalization is not based on the sun – the only global force that is equally available to all and whose bounty is so great that it need never be fully tapped. Only with solar in place of fossil energy can the world reach the pinnacle of its potential. As long as economic progress depends on resources found only in a few regions, there will inevitably be increasingly bitter conflicts in which national interests will come before the interests of the planet, national economies before the economy as a whole, short-term before long-term interests and individuals and companies before society. The global hierarchies that have grown and continue to grow out of fossil energy supplies stand in the way of a new era in which people can make as close to an independent living as can be achieved, and in which people can make their contribution to global output according to the measure of their ability and need. The existing hierarchies, however, are ironing out economic and cultural differences, depriving the world of its vibrant diversity. Cultural destitution is following hard on the heels of its economic twin.

It is because the global flow of fossil resources has for a long time been widening the scope of possibility and opportunity for increasing numbers of people in the industrialized nations that people now fail to see that the same resource flow now has the opposite effect, narrowing the range of opportunities for increasing numbers of people, and ultimately for everybody. Global resource conflict, environmental catastrophe, fossil energy prices that are unaffordable for most of the world's population, the economic crises to come as supplies dwindle — all these put the world in grave danger of turning back the clock. Hard-won achievements of civilization may be lost: the UN and international law, international treaties, the global economy itself. The most likely consequence of the struggle to control dwindling fossil reserves is a deep decline in the global economy, leading ultimately to the fall of global civilization itself.

The solar global economy makes possible a new global division of labour. Each national economy exploits the resources directly afforded them by the sun, resources that no-one can take away; all other needs are satisfied by the free interaction of supply and demand. Only in this way can the rich diversity of global culture be maintained and revitalized, or further developed through mutual enrichment.

The globalization process is a roller-coaster ride driven by fossil fuels. The faster it goes, the more frightening and bruising a ride it is for human passengers and the natural world alike. Dwindling numbers of people are able to climb aboard, while growing proportions are tossed out of the carriage. By contrast, the new division of labour in the solar global economy to come encompasses a whole variety of swings and round-abouts, some small, some large, all offering a much calmer ride, much less violent to — and more under the control of — their passengers. There will always be room for more attractions, with plenty of space for all comers. The solar global economy affords much greater freedom and scope for the productive use of technology because of the countless individual practical applications that, in combination with the immediate availability of the sun's power, it makes possible. Technology will no longer be the preserve of the few, who use it to impose technocratic constraints on everybody. The universal accessibility of technology will open the floodgates for many more new ideas and innovative applications. Growing numbers of independent producers and more diverse resource use will give rise to a whole range of new products. The solar global economy is an economy that does not wantonly destroy its resources, and which is thus free of constraints on its development.

By taking hold of the visible hand of the sun and producing from sustainable resources, the world remains close to the land, and its inhabitants meet in a freer and more just environment. From riches for the few, be they individuals, companies or societies, will increasingly come wealth for all, more justly and more equally distributed. Renewable resources will bring a new era of wealth-creating economic development, initiated not by bureaucratic fiat, but by the free choices of individuals.

References

Scenario

1   Arthur Koestler: *The Ghost in the Machine*. New York: The Macmillan Company 1968
2   Frederick Soddy: *Matter and Energy*. 1912, cited in Gerald Foley: *The Energy Question*. London: Penguin Books 1989
3   P S Dasgupta and G M Heal: *Economic Theory and Exhaustible Resources*. Cambridge: Cambridge Economic Handbooks 1995, pp469 et seq
4   Jürgen Habermas: *Technik und Wissenschaft als Ideologie* (Science and Technology as Ideology). Frankfurt am Main: Suhrkamp Verlag 1968, p90
5   Hans Immler: *Welche Wirtschaft braucht die Natur?* (What Sort of Economy does Nature Need?) Frankfurt am Main: S Fischer 1993, p26
6   Friedrich Schmidt-Bleek: *Das MIPS-Konzept. Weniger Naturverbrauch – mehr Lebensqualität durch Faktor 10* (The MISU (Material Input per Service Unit) Concept: reduced consumption and enhanced quality of life with Factor 10). Munich: Droemer 1998, p55
7   Ernst Ulrich von Weizsäcker, Amory B Lovins and L Hunter Lovins: *Factor 4: doubling wealth – halving resource use. The new report to the Club of Rome*. London: Earthscan 1997
8   Hanns Maull: *Raw Material, Energy and Western Security*. London: Macmillan Press 1984
9   Hans Kronberger: *Blut für Öl. Der Kampf um die Ressourcen* (Blood for Oil: the battle for natural resources). Vienna: Uranus Verlag 1998
10 Michio Kaku: *Visions : how science will revolutionize the 21st century*. New York: Bantam Books 1998
11 Gregory Stock: *Metaman: the merging of humans and machines into a global superorganism*. New York: Simon & Schuster 1993
12 Jeremy Rifkin: *Biosphere Politics: a cultural odyssey from the Middle Ages to the New Age*. New York: Crown 1991
13 Wilhelm Fucks: *Formeln zur Macht. Prognosen über Völker, Wirtschaft, Potentiale* (The Formula for Power : the future of nations and economies). Stuttgart: Deutsche Verlags-Anstalt 1965
14 *Global 2000* (Report to President Carter). Washington, DC: United States Government Printing Office 1980
15 Werner Müller: *Des Feuers Macht* (The Power of Fire). Herne: Heitkamp 1986, pp110 et seq
16 The extract is taken from the English edition of 1888

17   Jan Ross: 'Stunde der Politik' (The Hour of the Politicians). *Die Zeit* 42 1998

18   David C Korten: *When Corporations Rule the World.* London: Earthscan 1995, pp119–181

19   Oskar Lafontaine and Christa Müller: *No Fear of Globalisation: welfare and work for all.* London: Verso Books, in preparation

20   Edward Goldsmith: *The Way: an ecological world view.* Dartington: Foxhole 1996, pp445 et seq

21   Francis Bacon: *New Atlantis* (first published 1627)

22   Victor C Ferkiss: *Nature, Technology, and Society: the cultural roots of the current environmental crisis.* New York: New York University Press 1993, pp105–118

23   François Quesnay: *Maximes générales du gouvernement économique d'un royaume agricole* (General Principles for the Economic Administ-ration of an Agricultural Kingdom) (first published 1758)

24   Hermann Scheer: *Zurück zur Politik* (Back to Politics). Bochum: Verlag Ponte Press 1998 (2nd edition), pp197 et seq

25   Hermann Scheer: *A Solar Manifesto: the need for a total solar energy supply – and how to achieve it.* London: James & James 1994, 2nd edition 2001

26   Carl Amery: *Die Botschaft des Jahrtausends: Von Leben, Tod und Würde* (Message for the Millennium: of life, death and dignity). Munich: List-Verlag 1994, p177

27   European Commission: Communication from the Commission. *Energy for the Future: renewable sources of energy. White Paper for a community strategy and action plan.* 1997, document number COM (97) 599 final. Brussels: European Commission

28   Hermann Fischer and Hanswerner Mackwitz: 'Erneuerbare Rohstoffe, Solare Materialien : das zweite Schlüsselelement einer solaren Alternative' (Renewable Materials, Solar Materials: the second key element of a solar alternative). *Solarzeitalter* no 21 1995, pp31 et seq

29   Carl Amery: *Hitler als Vorläufer. Auschwitz : der Beginn des 21 Jahrhunderts* (Foretaste of the Future: Hitler, Auschwitz and the beginning of the 21st century). Munich: Luchterhand Literaturverlag 1998

30   Donald Worster: *The Wealth of Nature: environmental history and the ecological imagination.* New York/Oxford: Oxford University Press 1993

Part I

Chapter 1

1   Jean-Claude Debeir, Jean-Paul Deleage and Daniel Hémery: *In the Servitude of Power: energy and civilisation through the ages.* London/ Atlantic Highlands, New Jersey: Zed Books 1991

2   Vaclav Smil: *Energy in World History.* Boulder, Colorado: Westview Press 1994

3   Ralf-Peter Sieferle: *Der unterirdische Wald. Energiekrise und industrielle Revolution* (The Subterranean Forest: energy crises and industrial revolution). Munich: C H Beck 1982
4   World Resources Institute: *Resource Flows: the material basis of industrial economies*. Washington, DC: World Resources Institute 1997, pp33 et seq
5   European Commission 1997 (see Scenario, note 27)
6   Ugo Bilardo and Giuseppe Mureddu: *Energy, Raw Materials for Industry and International Cooperation*. Rome: ENEL 1989, p26
7   Helge Hveem: 'Minerals as a Factor in Strategic Policy and Action'. In: Arthur H Westing: *Global Resources and International Conflict*. Oxford: Oxford University Press 1986, p61
8   Bilardo and Mureddu 1989 (see note 6), p217
9   Energy and Defense Project: *Dispersed, Decentralized and Renewable Energy Sources: alternatives to national vulnerability and war*. Washington, DC: Federal Emergency Management Agency (FEMA) December 1980, p6
10   Anthony Sampson: *The Seven Sisters: the great oil companies and the world they made*. Dunton Green: Hodder and Stoughton 1993
11   Markus Timmler: 'Was UNCTAD ist und sein könnte' (What UNCTAD Is and Could Be). *Orientierungen zur Wirtschafts- und Gesellschaftspolitik* no 33 March 1987, pp55 et seq
12   Mohssen Massarrat: *Endlichkeit der Natur und Überfluß in der Marktökonomie* (The Limitations of Nature and Affluence in Market Economies). Marburg: Metropolis-Verlag 1993
13   K Ballod-Atlanticus: *Der Zukunftsstaat. Produktion und Konsum im Sozialstaat* (The Future State: production and consumption in the welfare state). Stuttgart: Dietz 1920
14   Nikolaus Eckardt, Margitta Meinerzhagen and Ulrich Jochimsen: *Die Stromdiktatur. Von Hitler ermächtigt : bis heute ungebrochen* (The Electricity Dictatorship : enabled by Hitler and unbroken to this day). Hamburg: Rasch und Röhring 1985
15   Thomas B Hughes: *Networks of Power: electrification in western society*. Baltimore/London: Johns Hopkins University Press 1983
16   Eugene D Cross: *Electricity Utility Regulation in the European Union*. Chichester: Wiley 1996, pp21 et seq
17   Lutz Metz and Rainer Osnowski: *RWE. Ein Riese mit Ausstrahlung* (RWE : an influencial giant). Cologne: Kiepenheuer und Witsch 1996
18   Daniel M Berman and John T O'Connor: *Who Owns the Sun?*. White River Junction: Chelsea Green 1996, p65
19   Oleg W Britwin: 'Die Reform der russischen Stromwirtschaft : neue Möglichkeiten der Kooperation' (Reforming the Russian Energy Sector: new horizons for cooperation). In: Andrei Kuxenko/Friedemann Müller (eds): *Deutsch-russische Energiekooperation unter Globalisierungsdruck* (Russo-German Energy-Sector Cooperation in the Globalised Market). Ebenhausen: Stiftung Wissenschaft und Politik 1998, pp39 et seq

20 Attilio Bisio, Sharon Boots and Paula Siegel (eds): *The Wiley Encyclopedia of Energy and the Environment.* New York: Wiley 1997, p1262
21 Paul Ehrlich, Ann Ehrlich and John Holdren: *Ecoscience.* San Francisco: W H Freeman 1979, p416
22 Bilardo and Mureddu 1989 (see note 6)
23 World Energy Council: *Energy for Tomorrow's World.* New York: St Martin's Press 1993, p102
24 ifo Institut für Wirtschaftsforschung: *ifo Investitionsberichte: Erhebung 1997/98. Öffentliche Elektrizitätsversorgung* (ifo Investment Reports 1997/98: electricity supplies). Munich: ifo 1998
25 ifo Insitut für Wirtschaftsforschung: *ifo Schnelldienst* no 32 1998, p13
26 Armin Witt: *Unterdrückte Entdeckungen und Erfindungen* (Supressed Inventions and Discoveries). Frankfurt: Ullstein-Verlag 1993
27 European Commission: Working Document of the Commission: Summary of the Results of the Public Consultation on the Green Paper on the Convergence of the Telecommunications, Media and Information Technology Sectors; Areas for Further Reflection. Document reference: SEC (98) 1284

Chapter 2

1   Scheer 2001 (see Scenario, note 25), pp89 et seq
2   Kulsum Ahmed: *Renewable Energy Technologies: on status and costs of selected technologies.* World Bank Technical Paper no 240 (Energy Series). Washington, DC: World Bank 1994
3   David O Hall and Frank Rosillo-Calle: 'Biomass: a future renewable carbon feedstock for energy'. In: V N Parman, H Tributsch, A Bridgwater and D O Hall: *Chemistry for the Energy Future.* Oxford: Blackwell Science 1999, pp101 et seq, 109, 118
4   A Strehler: 'Energie aus Biomasse' (Energy from Biomass). *Energie-Dialog* no 3/4 1991
5   Hall and Rosillo-Calle 1999 (see note 3)
6   Harry Lehmann and Thorsten Reetz: *Zukunftsenergien. Strategien einer neuen Energiepolitik* (Future Energies: strategies for a new energy policy). Berlin: Birkenhäuser Verlag 1995
7   Wolfgang Voigt: 'Weltmacht Atlantropa' (The Transatlantic Superpower). *Die Zeit* 23 1991
8   Yukinori Kuwano: 'Photovoltaic Electricity: an industrial perspective'. In: Helmar Krupp: *Energy Politics and Schumpeter Dynamics.* Tokyo: Springer-Verlag 1992, pp202–205
9   'Solar Energy and Space'. World Summit. UNESCO: Paris 1993
10 Franz Alt: *Die Sonne schickt uns keine Rechnung* (The sun sends no bills). Munich: Piper 1994

Part II

Chapter 3

1   Samuel P Huntington: *The Clash of Civilizations and the Remaking of World Order*. New York: Simon & Schuster 1996

2   International Energy Agency: *World Energy Outlook 1998*. Paris: International Energy Agency 1998

3   Stockholm Environment Institute: *Global Energy in the 21st Century: patterns, projections and problems*. Stockholm: Stockholm Environment Institute 1995, p29

4   Global 2000 (see Scenario, note 14), pp387 et seq

5   Jörg Schindler and Werner Zittel: 'Wie lange reicht das billige Öl?' (How Long will Cheap Oil Last?). *Scheidewege. Jahresschrift für skeptisches Denken* (Crossroads. Journal for Sceptical Thinking) 1998–1999, pp320 et seq

6   Bundesanstalt für Geowissenschaften und Rohstoffe (BGR), unpublished paper. Hanover: BGR 1999

7   Colin J Campell: *The Coming Oil Crises*. Petroconsultants: Geneva 1997

8   Günter Eickhoff and Hilmer Rempel: 'Weltreserven und ressourcen beim Energierohstoff Erdgas' (Global Reserves and Resources for the Energy Source Natural Gas). *Energiewirtschaftliche Tagesfragen* (Topics in Energy Economics) 1995, pp709 et seq

9   International Energy Agency: *Coal Information*. Paris: International Energy Agency 1993, pp138 et seq

10 World Energy Council: *Energy for Tomorrow's World*. New York: St Martin's Press 1998, p90

11 Jochen Diekmann et al: 'Fossile Energieträger und erneuerbare Energiequellen' (Fossil Fuels and Renewable Energy Sources). *Monographien des Forschungszentrum Jülich* vol 25 1995, p113

12 Klaus Traube and Hermann Scheer: 'Kernspaltung, Kernfusion, Sonnenenergie : Stadien eines Lernprozesses' (Nuclear Fission, Nuclear Fusion, Solar Energy : stages of a learning process). *Solarzeitalter* no 2 1998, p22 et seq

13 Friedrich-Wilhelm Wellmer: 'Lebensdauer und Verfügbarkeit mineralischer Rohstoffe' (Duration and Availability of Mineral Resources). In: J Zemann (ed): *Energievorräte und mineralische Rohstoffe: wie lange noch?* (Energy Supplies and Mineral Resources: How much longer?). Vienna: Verlag der Österreichischen Akademie der Wissenschaften 1998, pp47 et seq

14 F William Engdahl: *Mit der Ölwaffe zur Weltmacht : Der Weg zur neuen Weltordnung* (Playing the Oil Card to Become a Superpower: the road to a new world order). Wiesbaden: Böttiger Verlag 1997 (3rd edition), pp244 et seq

15 Gabriele Venzky: 'Kampf ums Öl' (The Struggle for Oil). *Die Zeit* 43 1998

16 Bartholomäus Grill and Caroline Dumay: 'Der Söldner-Konzern' (The Mercenary Corporation). *Die Zeit* 4 1997
17 Zbigniew Brzezinski: *The Grand Chessboard: American primacy and its geostrategic imperatives.* New York: Basic Books 1997
18 'Apocalypse soon in Minerals'. *Defense Week* 22 September 1980
19 *Donald E Fink:* 'Availability of Strategic Materials'. *Aviation Week and Space Technology* 5 May 1990
20 Gernot Erler: *Global Monopoly. Weltpolitik nach dem Ende der Sowjetunion* (Global Monopoly: World politics following the collapse of the Soviet Union). Berlin: Aufbau Taschenbuch Verlag 1998, pp86 et seq
21 Kronberger 1998 (see Scenario, note 9), pp141 et seq
22 All data taken from: International Energy Agency: *Energy Statistics and Balances of Non-OECD Countries.* Paris: International Energy Agency 1998
23 Günter Mohrmann: 'Der Kampf um die knappen Wasserressourcen. Krieg um knappes Wasser?' (The Struggle for Scarce Water Supplies. A war fought for water?). In: Hans Krech: *Vom II. Golfkrieg zur Golf-Friedenskonferenz* (From the Second Gulf War to the Gulf Peace Conference). Bremen: Edition Temmen 1996, pp367 et seq
24 Fritz R Glunk (ed): *Das MAI und die Herrschaft der Konzerne* (The MAI and Corporate Domination). Munich: Deutscher Taschenbuch Verlag 1998
25 William D Heffernan and Douglas H Constance: 'Transnational Corporation and the Globalization of the Food System'. In: Alessandro Bonanno et al: *From Columbus to Con Agra.* Kansas City: University Press of Kansas 1994, pp29 et seq
26 Harald Maass: 'Ein mächtiger Sturm kündigt sich leise an' (Silent Heralds of a Mighty Storm). *Frankfurter Rundschau* 16 March 1999
27 Samir Amin: *Empire of Chaos.* New York: Monthly Review Press 1992

Chapter 4

1 Rolf Peter Sieferle: 'Energie' (Energy). In: Franz-Josef Brüggemeier and Thomas Rommelspacher: *Besiegte Natur* (Nature Conquered). Munich: C H Beck 1987, p32
2 UNCHS (Habitat): *An Urbanizing World: global report on human settlements.* Oxford: Oxford University Press 1996
3 Dieter Schott: *Energie und Stadt in Europa. Von der vorindustriellen Holznot« bis zur Ölkrise der 70er Jahre* (Energy and Cities in Europe: From the Pre-industrial Wood Shortage to the Oil Shock of the 70s). Stuttgart: F Steiner 1997, p31
4 Smil 1994 (see Chapter 1, note 2), pp208 et seq
5 Otto Königer: *Die Konstruktionen in Eisen* (Building with Iron). Leipzig: Gerhardts Verlag 1902 (new edition, Hanover 1993)
6 Helmut Tributsch: 'Wohnen mit der Sonne' (Living with the Sun). *Solarzeitalter* no 2 1991, pp22–30

7   Sophie Behling and Stefan Behling: *Sol Power: The Evolution of Solar Architecture*. Munich: Prestel-Verlag 1996

8   Le Corbusier: *The Athens Charter*. New York: Grossman Publishers 1973

9   Michael Mönninger: 'Stadt statt Staat. Illusionen von Global- und Mega-Cities' (City, not State: The illusions of global and megacities). *IZ3W* May/June 1999, pp21 et seq

10 Hartmut Häußermann: 'Die selbstzerstörerische Sehnsucht nach der Global City' (The Self-Destructive Yearning for the Global City). *Frankfurter Rundschau* 22 May 1999

11 Michael Knoll and Rolf Kreibich: *Solar City. Sonnenenergie für die lebenswerte Stadt* (Solar City: Solar energy for healthy cities). Weinheim: Beltz-Verlag 1992

12 *Cities for the Future: dream or nightmare?* Panos Briefing no 34. London: Panos June 1999, pp4 et seq

13 Idriss Jazairy, Mohiuddin Alamgir and Theresa Panuccio: *The State of World Rural Poverty*. London: IT Publications 1992, p2

14 Birgit Rheims: 'Migration und Flucht' (Migration and Refugees). In: Ingomar Hauchler, Dirk Messner and Franz Nuschler: *Globale Trends 1998* (Global Trends 1998). Frankfurt am Main: Fischer Taschenbuch Verlag 1997, pp97 et seq

15 Axelle Kabou: *Et si l'Afrique refusait le développement?* (What if Africa Says No to Development Aid?). Paris: L'Harmattan 1991

16 Winrich Kühne: *The Great Lakes Region Crisis: Improving African and international capabilities for preventing and resolving violent conflicts*. Ebenhausen: Stiftung Wissenschaft und Politik 1997

17 Walt Rostow: *The Stages of Economic Growth: a non-communist manifesto*. 1960 (3rd edition Cambridge: Cambridge University Press 1994)

18 Scheer 2001 (see Scenario, note 25), pp41 et seq

19 Chengdu Biogas Research Institute: *The Biogas Technology in China*. Beijing: China Agricultural Publishing House 1989

20 Peter Oesterdiekhoff: *Dimensionen der Energiekrise in Afrika südlich der Sahara* (The Extent of the Energy Crisis in sub-Saharan Africa). Energy and Development Discussion Papers Series. Bremen: Institut für Weltwirtschaft und Internationales Management 1991, pp12 et seq

21 L Mohapeloa: 'The Case of Lesotho'. In: M R Bhagavan and S Karekezi: *Energy Management in Africa*. London: Zed Books 1992, pp9 et seq

22 Cheryl Payer: *The World Bank: a critical analysis*. New York: Monthly Review Press 1982, pp101 et seq

23 Michael Tanzer: *The Energy Crisis*. New York: Monthly Review Press 1974, p108

24 Payer 1982 (see note 22), pp186 et seq

25 Daphne Wysham: *The World Bank and the G7: changing the earth's climate for business*. Washington, DC: Institute for Policy Studies 1997

26 Anil Cabraal, Mac Cosgrove-Davies and Loretta Schaeffer: *Best Practices for Photovoltaic Household Electrification Programs*. World Bank Technical Paper no 324. Washington, DC: World Bank 1996

27 EUREC Agency, WIP Munich, EUROSOLAR et al: *PV for World's Villages: Final Report*. November 1996

28 Peter Meyns: 'From Coordination to Integration. Institutional Aspects of the Development of SADC'. In: Heribert Dieter (ed): *The Regionalization of the World Economy and Consequences for Southern Africa*. Marburg: Metropolis Verlag 1997, pp163 et seq

29 Brigitte Weidlich: 'Ein Energienetz vom Äquator bis zum Kap' (An Energy Grid from the Equator to the Cape). *Solarzeitalter* no 3 1998, pp29 et seq

30 Bilardo and Mureddu 1989 (see Chapter 1, note 6), p215

31 Payer 1982 (see note 22), p158. Quotation taken from Greg Lanning: *Africa Undermined*

Chapter 5

1   Roland Barthes: *Mythologies*. London: Vintage 1993 (first published 1972)

2   Wolfgang Palz: 'Renewable Energies in Europe: statistics and their problems'. *International Journal of Solar Energy* vol 17 1995, pp73 et seq

3   Stockholm Environment Institute 1995 (see Chapter 3, note 3), p24

4   Gustav R. Grob: *New Total Approach to Energy Statistics and Forecasting*. Geneva: Script World Sustainable Energy Coalition 1999

5   IEA 1998 (see Chapter 3, note 2)

6   Shell: 'Energy for Developments'. Selected Paper. London: Shell International Petroleum Company, Shell Centre 1994

7   Irm Pontenagel: *Das Potential erneuerbarer Energien in der Europäischen Union. Ansätze zur Mobilisierung erneuerbarer Energien bis zum Jahre 2020* (The Potential for Renewable Energy in the European Union: proposals for mobilizing renewable energy towards 2020). Heidelberg: Springer-Verlag 1995

8   Scheer 2001 (see Scenario, note 25), pp89 et seq

9   Scheer 2001 (see Scenario, note 25), pp76 et seq

10 Scheer 2001 (see Scenario, note 25), pp48 et seq

11 Greenpeace: *Energy Subsidies in Europe*. Amsterdam: Greenpeace International 1997

12 Richard Douthwaite: 'The Growth Illusion'. In: Steven Gorelick: *Small is Beautiful, Big is Subsidised*. Devonshire: International Society for Ecology and Culture 1998, pp23 et seq

13 Scheer 2001 (see Scenario, note 25), pp230 et seq

14 Gorelick 1998 (see note 12)

15 Douthwaite 1998 (see note 12)

16 Douthwaite 1998 (see note 12)

17 A I C N Reddy, R H William and T B Johansson: *Energy after Rio: prospects and challenges.* New York: United Nations Development Programme 1997, p137
18 Gorelick 1998 (see note 12)
19 Stockholm Environment Institute 1995 (see Chapter 3, note 3), pp39 et seq
20 Umweltbundesamt (Federal Environment Ministry): *Ökologische Bilanz von Rapsöl als Ersatz von Dieselkraftstoff* (Environmental Cost–Benefit Analysis of Rapeseed Oil as a Subsitute for Diesel Oil). Berlin: Umweltbundesamt 1993
21 Scheer 2001 (see Scenario, note 25), pp121 et seq
22 Witt 1993 (see Chapter 1, note 26)
23 Ralf Bischof: 'Wenn mit Energiedichten gegen erneuerbare Energie angedichtet wird' (Using Energy Densities to Make Renewable Energy Sound Dense). *Solarzeitalter* no 3 1994, pp10 et seq
24 Klaus Heinloth: *Die Energiefrage* (The Energy Question). Wiesbaden: Vieweg 1997, pp336 et seq
25 Deutsche Physikalische Gesellschaft: 'Energie memorandum 1995 der Deutschen Physikalischen Gesellschaft. Zukünftige klimaverträgliche Energienutzung und politischer Handlungsbedarf zur Markteinführung neuer emissionsmindernder Techniken' (Future Climate-neutral Energy Use and the Need for Political Action to Bring New Emissions-reducing Technologies onto the Market). *Physikalische Blätter* vol 51 1995, pp388 et seq
26 Federico Di Trocchio: *Newtons Koffer: Geniale Außenseiter, die die Wissenschaft blamierten* (Newton's Suitcase: brilliant outsiders who showed science up). Frankfurt am Main: Campus Sachbuch 1998, pp244 et seq
27 Viktor Gorgé: *Philosophie und Physik* (Philosophy and Physics). Berlin: Duncker 1960, p128
28 Thure von Uexküll: *Der Mensch und die Natur. Grundzüge einer Naturphilosophie* (Man and Nature: Foundations for a Natural Philosophy). Bern, Francke 1953, pp46 et seq
29 Joachim Radkau: *Aufstieg und Krise der deutschen Atomwirtschaft* (The Rise of the German Nuclear Sector and its Crisis). Reinbek bei Hamburg: Rowohlt rororo 1983, pp462 et seq
30 Franz Josef Radermacher: *Information Society, Globalisation and Sustainable Development.* Ulm: Forschungsinstitut für anwendungsorientierte Wissensverarbeitung (FAW) Ulm 1998

Part III

1   James M Utterback: *Mastering the Dynamics of Innovation.* Boston, Mass: HBS Press 1994, p231

Chapter 6

1 Berman and O'Connor 1996 (see Chapter 1, note 18), pp12 et seq
2 Wolfgang Wismeth (ed): *Photovoltaik Handbuch III* (GWU Solar GmbH Handbook of Photovoltaics III). Fürth: GWU Solar GmbH 1997
3 'Von der Armbanduhr bis zum Taschenrechner: Wo die Sonne die Batterie ersetzt' (From the Wristwatch to the Pocket Calculator: where the sun replaces batteries). *Photon* no 1 1998, pp46–49
4 *Financial Times* June 1999
5 Berman and O'Connor 1996 (see Chapter 1, note 18), pp203 et seq
6 German Environment Ministry: *UBA-Texte* (Environment Ministry Papers) no 45/97 1997
7 Chihiro Watanabe: *Industrial Dynamism and the Creation of a 'Virtuous Circle': the case of photovoltaic power generation development in Japan.* Unpublished paper, May 1999
8 Statistics supplied by Gesellschaft für innovative Energieumwandlung und -speicherung (EUS): Gelsenkirchen 1999
9 Michael Shnayerson: *The Car that Could.* New York: Random House 1996, pp262 et seq
10 Rudolf Weber: *Der sauberste Brennstoff* (The Cleanest Fuel). Oberbözberg: Olynthus Verlag 1988
11 Jürgen Kleinwächter: *Das Sepak-System* (The Sepak System). Lörrach 1990
12 Martin Werdich and Kuno Kübler: *Stirling-Maschinen. Grundlagen, Technik, Anwendungen* (Stirling Engines: Principles, Technologies, Applications). Staufen: Ökobuch Verlag 1999 (7th edition)
13 Volker Schindler: *Kraftstoffe für morgen. Eine Analyse von Zusammenhängen und Handlungsoptionen.* (Fuels for Tomorrow: an analysis of the background and the options). Berlin: Springer-Verlag 1997, pp59 et seq
14 Peter Winkelkötter: *Aufbereitungsanlage für Biomasse* (Biomass Processing Plant); Waste Energy Action: *Conversion Plant for Biomass.* Singapore
15 Rudolf Nußstein: 'Einsatz von Alkohol als Wasserstoffspeicher für Brennstoffzellenantriebe in Kraftfahrzeugen' (Alcohol as a Hydrogen Storage Medium for Fuel Cell Engines in Motor-Cars). *Solarzeitalter* no 1 1999, pp21 et seq
16 Konrad Scheffer: 'Der Landwirt als Energiewirt: Anbau- und Nutzungskonzepte von Biomasse' (The Farmer as Energy Producer: proposals for the production and application of biomass). *Solarzeitalter* no 1 1999, pp26 et seq
17 Volkswagen AG/ZSW: *Renewable Methanol from Hydrogen for Road Transformation*
18 Bodo Wolf: 'Biobenzin. Ein Beitrag zur Weiterentwicklung der Energiewirtschaft' (Bio-Petrol: a contribution towards the further development of the energy sector). Lecture given at the Eurosolar Conference 'Der Landwirt als Energiewirt' (Farmers as Energy

Producers) during the Grüne Woche Berlin exhibition, 23 January 1999

19   European Commission: *Biofuels. Application of Biologically Derived Products as Fuels or Additives in Combustion Engines.* Document reference: EUR 15647 EN 1994

20   Klaus Daniels: *Low Tech. Light Tech. High Tech. Bauen in der Informationsgesellschaft* (Low Tech, Light Tech, High Tech: construction in the information society). Basel: Princeton Architectural Press 1998, p9

21   Thomas Herzog: *Solar Energy in Architecture and Urban Planning.* Munich: Prestel Publishing 1996

22   Tributsch 1991 (see Chapter 4, note 6)

23   Sir Norman Foster and Hermann Scheer: *Solar Energy in Architecture and Urban Planning. Proceedings of the Third European Conference on Architecture.* Felmersham, UK: James & James 1993, p18

24   Radermacher 1998 (see Chapter 5, note 30)

25   John Naisbitt: *Global Paradox.* New York: William Morrow 1994, pp49–51

26   Philippe Quéau: 'Auf der Überholspur der Datenautobahn' (In the Fast Lane of the Information Superhighway). *Le Monde Diplomatique* (reproduced in German in: *die tageszeitung*) 2 February 1999, p24

27   Helmut Tributsch: 'Programm für Solarenergieforschung und –technologien' (Programme for Solar Energy Research and Solar Technologies). *Solarzeitalter* no 2 1999, pp6 et seq

Chapter 7

1   Carsten Alsen and Ottmar Wassermann: *Die gesellschaftliche Relevanz der Umwelttoxikologie* (The Social Relevance of Environmental Toxicology). IIUL Report. Berlin: Wissenschaftszentrum 1986, p6

2   Heinrich Reitz: 'Bestimmungsgründe für den Einsatz nachwachsender Rohstoffe in der chemischen Industrie' (The Case for Renewable Materials in the Chemicals Industry). In: *Nachwachsende Rohstoffe : von der Forschung zum Markt* (Renewable Materials: from research to marketplace). Series Gülzower Fachgespräche. Fachagentur Nachwachsende Rohstoffe (www.fnr.de/veroff/fnr_gf06.pdf) 1998, p20

3   Karl-Otto Henseling: *Ein Planet wird vergiftet: Der Siegeszug der Chemie: Geschichte einer Fallentwicklung* (A Planet is Poisoned. the triumph of chemistry: a case history). Reinbek bei Hamburg: Rowohlt 1992

4   G Grassi: 'Job Potential of Biomass'. In: Chartier et al (eds): *Proceedings of the 9th European Bioenergy Conference: Volume 1.* Oxford: Elsevier Science 1996, pp419 et seq

5   Hermann Fischer: *Plädoyer für eine sanfte Chemie: Über den nachhaltigen Gebrauch der Stoffe* (The Case for Soft Chemistry: on the sustainable use of materials). Karlsruhe: C F Müller Verlag 1993, p57

6   Fischer 1993 (see note 5)
7   Fischer 1993 (see note 5)
8   KATALYSE Institut für angewandte umweltforschung: *Leitfaden Nachwachsende Rohstoffe* (Handbook of Renewable Resources). Heidelberg: C F Müller Verlag 1998
9   Dirk Schäfer: 'Einsatz und Potential naturfaserverstärkte Kunststoffe in der Automobilindustrie' (The Use and Potential of Natural Fibre Reinforced Plastics in the Car Industry). In: *Nachwachsende Rohstoffe* 1998 (see note 2), p48
10 Fischer 1993 (see note 5)
11 Theo Mang: 'Zielgruppenanalyse für biologisch abbaubare Schmierstoffe' (Target Group Analysis for Bio-Degradable Lubicants). In: *Nachwachsende Rohstoffe* 1998 (see note 2), p70
12 Götz Harnischfeger: 'Marktpotentiale von Arzneimitteln aus pflanzlichen Werkstoffen' (Market Potential of Pharmaceuticals derived from Plant Materials). In: *Nachwachsende Rohstoffe* 1998 (see note 2)
13 Henkel KgaA: *Umwelt, Sicherheit, Gesundheit. Daten und Fakten für 1998.* (Environment, Safety, Health. Data and Facts for 1998). Düsseldorf: Henkel KgaA 1998
14 Fischer 1993 (see note 5), pp66 eq seq
15 Report of the German Parliamentary Commission of Inquiry 'Gestaltung der technischen Entwicklung, Technikfolgen-Abschätzung und Bewertung' (Shaping Technological Development, Evaluating and Assessing the Consequences of Technologies). *Nachwachsende Rohstoffe* 23/90 1990, pp47 et seq
16 *Chemielexikon* (Dictionary of Chemistry) vol 1. Stuttgart: Georg Thieme Verlag 1989, p426
17 Abdel Mottaleb: *Proceedings of the 4th International Conference on Solar Energy Storage and Applied Photochemistry.* Cairo January 1997
18 United Nations Environment Programme (UNEP): *Global Biodiversity Assessment.* Cambridge: Cambridge University Press 1995
19 Wolfgang Franke: *Nutzpflanzenkunde* (Crop Science). Stuttgart: Georg Thieme Verlag 1989
20 Jack Herer. *Die Wiederentdeckung der Nutzpflanze Hanf* (The Rediscovery of Hemp). Munich: Heyne 1994
21 Bundesanstalt für Geowissenschaften und Rohstoffe 1999 (see Chapter 3, note 6), p18
22 Hall and Rosillo-Calle 1999 (see Chapter 2, note 3)
23 H Lundquist: *Whole Tree Harvesting: ecological consequences.* Maritimes Region Information Report M-X-191. Canada 1994
24 Daniel Querol: *Genetic Resources: our forgotten treasure.* London: Zed Books 1992, pp26 et seq
25 Fischer 1993 (see note 5), p122
26 José Lutzenberger: 'Die Absurdität der modernen Landwirtschaft' (The Absurdity of Modern Agriculture). *Solarzeitalter* no 3 1999, pp3 et seq

27   Vandana Shiva: *Biotechnology and the Environment*. Pol Pinong: Third
     World Network pp34 et seq
28   José Lutzenberger 1999 (see note 26)
29   Vandana Shiva: *Biopiracy: the plunder of nature and knowledge*. New Delhi:
     South End Press 1998, pp43 et seq
30   Ricarda A Steinbrecher and Pat Roy Mooney: 'Terminator Technology:
     the threat to world food security'. *The Ecologist* vol 28 no 5 1998,
     pp276 et seq
31   Ulrich Dolata: *Politische Ökonomie der Gentechnik* (The Political
     Economics of Genetic Manipulation). Berlin: edition sigma 1996,
     pp183 et seq
32   *Aufbruchstimmung 1998. Report über die Biotechnologie-Industrie in Deutschland*
     (1998 Optimism. Report on the Biotech Industry in Germany).
     Stuttgart: Ernst & Young 1998, p14

Chapter 8

1   Peter Hennicke: *Least-Cost Planning als Element einer Einsparstrategie: Konzept
    und Erfahrungen in der Bundesrepublik.* (Least-Cost Planning as an
    Austerity Measure: concept and experiences in the FRG). Wuppertal
    1993
2   German Parliamentary Commission of Enquiry 1990 (see Chapter 7,
    note 15), p50
3   Rainer von Oheimb et al: *Energie und Agrarwirtschaft* (Energy and
    Agriculture). Münster 1987, pp40 et seq
4   Jules N Pretty: *Regenerating Agriculture: policies and practice for sustainability
    and self-reliance.* London: Earthscan 1995

Part IV

1   Dankwart Guratzsch: 'Naturkatastrophen häufen sich dramatisch'
    (Dramatic Increase in the Frequency of Natural Disasters). *Die Welt*
    25 June 1999
2   Scheer 1994 (see Scenario, note 25), pp210 et seq

Chapter 9

1   Eurosolar: *Solar Power Boats for Venice. study for the European Commission.*
    Bonn: Eurosolar 1995
2   Hermann Scheer: 'Das deutsche 100.000-Dächer-Photovoltaik-
    Programm' (The 100, 000 Roofs Photovoltaics Programme in
    Germany). *Solarzeitalter* no 4 1998, pp1 et seq
3   Irm Pontenagel: 'Solarenergie bricht Schweizer Parteifronten auf'
    (Solar Energy Crosses Party Lines in Switzerland). *Solarzeitalter* no 2
    1997, pp1 et seq

4 Timothy J Brennan et al.: *A Shock to the System: restructuring America's electricity industry.* Washington, DC: Resources for the Future 1996
5 Eurosolar: *EURERULE: Rechtliche Rahmenbedingungen für Erneuerbare Energien in der EU* (EURERULE: The legal framework for renewable energy in the EU.) Bonn: Eurosolar 1999
6 European Commission 1997 (see Scenario, note 27)
7 Rothe Report to the European Parliament. Document reference: PE 225-891/1/32. 7 May 1998
8 Hermann Scheer: 'EU-Einspeiserichtlinie und Einspeisegesetze für Erneuerbare Energien versus Einführungsquoten' (EU Laws and Directives on Electricity Supply to the National Grid versus Import Quotas). *Zeitschrift für neues Energierecht* no 2 1998, pp3 et seq
9 Eurosolar (ed): *Der Markt für Grünen Strom* (The Market for 'Green' Electricity) Bochum: Eurosolar 1998. (See eg Evert Jan Krajenbrink: 'Grüner Strom auf dem kalifornischen offenen Strommarkt' (Green Electricity on the Open Market in California), pp45 et seq
10 Infas Sozialforschung GmbH: 'Meinungen zu alternativen Energien' (Opinions on Alternative Energy Sources). *Solarzeitalter* no 1 1998, pp11 et seq
11 Reginald Scholz: 'Liberalisierung läßt die Preise fallen' (Liberalisation allows prices to fall). *Sonnenenergie & Wärmetechnik* no 2 1999, pp14 et seq
12 Ursula Sladek: 'Ein Irrweg. Ökostrom per Durchleitung gefährdet Einspeisegesetz' (On the Wrong Track: Green electricty via intermediaries jeopardizes grid feed-in act). *Neue Energie* no 4 1999, pp28 et seq
13 Klaus Traube: 'Umweltverbände zertifizieren Anbieter von Grünem Strom' (Certification of Green Electricity Suppliers by Environmental Organizations). *Solarzeitalter* no 2 1999, pp4 et seq
14 Heinz Ossenbrink: 'Energieleistung der Zukunft: Eine Chance für die erneuerbaren Energien' (Energy Generation in the Future: an opportunity for renewable energy). *Solarzeitalter* no 4 1998, pp6–10
15 Joseph A Schumpeter: *Capitalism, Socialism and Democracy.* New York: Harper & Brothers 1950 (3rd edition)

Chapter 10

1 Eurosolar: 'Entwicklung und Arbeitsplatzpotential Erneuerbarer Energien in der Europäischen Union' (The Development of Renewable Energy and its Job-Creation Potential in the EU). *Solarzeitalter* no 3 1997, pp11 et seq
2 Udo E Simonis: *Globale Umweltpolitik. Ansätze und Perspektiven* (Global Environment Policy: approaches and perspectives). Mannheim: Bibliographes Institut and F A Brockhaus 1996
3 German Parliament: *Antwort der Bundesregierung auf die Kleine Anfrage der Fraktion Bündnis 90/Die Grünen* (Government reply to question tabled by the Green Party). Paper 13/2156. Berlin: Bundestag 1995

4   Dennis R Henderson: 'Between the Farm Gate and the Dinner Plate:
     motivations for industrial change in the processed food sector'. In:
     Organisation for Economic Co-operation and Development: *The Future
     of Food*. Paris: OECD 1998, pp111 et seq
5   Johann Heinrich von Thünen: *Der isolierte Staat in Beziehung auf
     Landwirtschaft und Nationalökonomie* (The Isolated State with Reference
     to Agriculture and the National Economy). Aalen: Scientia Verlag
     1990 (first published 1826)
6   Julius K Nyerere: 'Sind universelle Sozialstandards möglich?' (Are
     Universal Social Standards Possible?). *epd-Entwicklungspolitik* no 12
     1998, pp37 et seq
7   Heribert Prantl: 'Wenn Bürger nicht mehr brav sein wollen' (When
     Citizens Disobey). *Süddeutsche Zeitung* 21–22 March 1998
8   Alfred Rest: 'Need for an International Court for the Environment'.
     Environmental Policy and Law 1994, pp173 et seq
9   Frank Biermann: 'Völkerrecht und Weltumweltpolitik' (International
     Law and Global Environment Policy). In: Udo E Simonis (eds):
     Weltumweltpolitik (Global Environment Policy). Berlin: edition sigma
     1996
10   Scheer 2001 (see Scenario, note 25), pp176 et seq
11   Ross Gelbspan: *The Heat is On: the high-stakes battle over earth's threatened
     climate*. Reading, Mass/Harlow: Addison-Wesley 1997

Chapter 11

1   Arran E Gare: *Postmodernism and the Environmental Crisis*. London:
     Routledge 1995, p34
2   Wilhelm Ostwald: *Die energetischen Grundlagen der Kulturwissenschaften*
     (The Energy Basis of the Humanities). Leipzig 1909, pp2 et seq
3   Jeremy Rifkin: *The End of Work: the decline of the global workforce and the dawn
     of the post-market era*. London: Penguin 2000
4   Mathias Greffrath: 'Freizeit, die sie meinen' (Freedom They Mean).
     Süddeutsche Zeitung 24 June 1998
5   Johano Strasser: *Wenn der Arbeitsgesellschaft die Arbeit ausgeht* (When the
     Working Society Runs Out of Work). Zürich: Pendo 1999
6   Elmar Altvater: *The Poverty of Nations: a guide to the debt crisis from Argentina
     to Zaire* London: Zed Books 1991

Index